Palaeontology

Proceedings of the
27th International Geological Congress

Stratigraphy	**Volume 1**
Palaeontology	**Volume 2**
Quaternary Geology and Geomorphology	**Volume 3**
Sedimentology	**Volume 4**
Precambrian Geology	**Volume 5**
Geology of Ocean Basins	**Volume 6**
Tectonics	**Volume 7**
Geophysics	**Volume 8**
Petrology (Igneous and Metamorphic Rocks)	**Volume 9**
Mineralogy	**Volume 10**
Geochemistry and Cosmochemistry	**Volume 11**
Metallogenesis and Mineral Ore Deposits	**Volume 12**
Oil and Gas Fields	**Volume 13**
Solid Fuel Mineral Deposits	**Volume 14**
Non-metallic Mineral Ores	**Volume 15**
Hydrogeology	**Volume 16**
Engineering Geology	**Volume 17**
Remote Sensing	**Volume 18**
Comparative Planetology	**Volume 19**
Mathematical Geology and Geological Information	**Volume 20**
History of Geology	**Volume 21**
Geological Education	**Volume 22**
Origin and History of Marginal and Inland Seas	**Volume 23**

Proceedings of the
27th International Geological Congress
Moscow 4–14 August 1984

Volume 2

PALAEONTOLOGY

VNU SCIENCE PRESS
Utrecht, The Netherlands
1984

VNU Science Press BV
P.O. Box 2073
3500 GB Utrecht
The Netherlands

First published 1984

ISBN 90–6764–011–5 Volume 2
ISBN 90–6764–009–3 set of 23 volumes

Printed in Great Britain by J. W. Arrowsmith Ltd, Bristol

PREFACE

The leading scientists in the many different fields of geology were invited by the Organizing Committee to present a paper on a specific problem in present-day geological science at the 27th International Geological Congress. The published proceedings of the Congress consist of twenty-three volumes. Each volume is dedicated to a particular aspect of geology. Together the volumes contain all of the contributions presented at the Congress.

The Organizing Committee is pleased to acknowledge the efforts of all of the participating scientists in helping to produce these proceedings.

Professor N. A. BOGDANOV
General-Secretary of the
Organizing Committee

CONTENTS

PALEONTOLOGICAL ASPECTS OF BIOMINERALIZATION

BARSKOV, I.S.
Moscow State University, USSR.

The biological mineralization, or formation of inorganic compounds (i.e. minerals, mineral aggregates, skeletons) by living organisms, is a regular stage in the evolution of living substance of the planet. The origin and development of biomineralization, as a product of integral interaction of living and inert matter is one of the most important mechanisms in the functioning of the biosphere.

Biomineralization researches are an integrated all-scientific problem, in which interests and methods of many scientific disciplines interlace. Paleontology and geological branches of science are of great importance in this investigation, because biomineralization originated in the inspected geological past, and skeletal paleontological records of life exist due to the process of biomineralization and they preserve the information about its origin and evolution for millions of years. Moreover, the understanding of biomineralization patterns make it possible to reconstruct fine physiological processes of the skeletogenesis of extinct organisms.

The study of compositional and structural evolution of mineral and organic components, which play the main role in the process of biomineralization and its morphological and microstructural expression in the form of skeletal parts of organisms and products of their functioning, belongs to the field of researches based on paleontological methods. From the paleontological point of view it seems significant to distinguish skeletal and nonskeletal biomineralization. This division states not only the difference in the function and mechanisms of biomineralization, but the

Proceedings of the 27th International
Geological Congress, Volume 2, pp. 1–12
PALAEONTOLOGY

existence of a very important stage in the evolution of organisms as well, that is the origination of integration between the two separate processes of skeletization and mineralization. There is no doubt, that the study of skeletal biomineralization is of main significance for paleontology, though in some cases it is difficult to delimit a "skeleton" and a "nonskeleton", and the nonskeletal biomineralization can be of great significance for geology and paleobiology, i.e. stromatolitogenesis and presence of biomineral grains in sediments.

1. <u>On the Evolution of Biomineralization Products Composition</u>

At present there are about forty minerals known to be formed biologically. Most of them are not non-skeletal and fulfil formative functions but various others including the metabolic ones.

One the whole, three groups of minerals are significant as skeletal, namely, the siliceous, calcium carbonates, calcium phosphates. Biomineralization of the two latter groups is a regular evolutionary stage and is determined by the availability of such biologically active elements as calcium and phosphorus. The evolutionary determination of the biomineralization based on silicon is less clear.

The siliceous biomineralization, or biosilicofication, means the formation of amorphous siliceous hydroxide, or opal. In the skeletal form it is characteristic of the most "lowly organized" groups, such as unicellular plants and animals (Diatomea, Silicoflagellata, Choanoflagellata, Radiolaria) and the parametazoan (Porifera). The mode of secretion is intracellular, unicellular.

The carbonate biomineralization or biocarbonatization, is represented by the formation of calcium carbonates (aragonite, calcite, vaterite). It is developed equally in all groups, except the radiolaria, being the main type of skeletal biomineralization in inverttebrates.

The mode of secretion is extracellular, unicellular or tissue. In the evolution of coelenterates, the calcite biomineralization in the Rugosa is supposed to be replaced by the aragonite one in the Scleractinia (13). On the contrary, in the bivalves and gastropod mollusca, taxa with calcite shells appeared geologically later than those with the aragonite shells.

The phosphate biomineralization or biophosphatization is represented in the skeletal form by the formation of calcium minerals of the apatite group, francolite, dallite, in the nonskeletal form by more metabolically active compounds. As for skeletogenesis, this biophosphatization is well-developed in the deuterostomian (inarticulate brachiopods, chordates). The secretion mode is tissue, extracellular, intercellular.

The biomineralization of other composition is rather restricted in skeletongenesis. The acantharians form a celestine skeleton, the mixed biomineralization with magnetite co-acting is found in chitonean jaws, temporal bones of birds; complex replacement of carbonate by phosphates is shown in mineralized dactilopodites in some crustacean.

The summary of distribution of skeletal mineralization types according to the composition among main phyla (Tab. 1) demonstrates the trend from the siliceous biomineralization to the phosphate one with the evolutionary advancement. This idea contradicts in a way the paleontological records of the skeletal composition of the Early Cambrian skeletal remains.

Lots of skeletal remains of the Early Cambrian organisms are of phosphate composition. This proves the mainly phosphate composition at the primary stages of skeletogenesis (8), as a reflection of the peculiarities of biochemical evolution (2).

However, some general speculations and facts indicate that the phosphate composition of skeletal remains of the Early Cambrian (Tommotian) organisms is a result of the secondary diagenetic phosphatization of primarily organic or slightly carbonatized covers. The problem of diagenetic phosphatization, no doubt, is to be studied specially, and at present one should carefully regard general biogeochemical conclusions based on the occurence of the fossils with phosphate skeleton.

2. On the Evolution of Skeletal Tissue Structure

The final morphological result of the biomineralization process is a structure, or a spatial organization and an orderly co-subjection of mineral and organic skeletal components. The study of skeletal of structure of ancient organisms is one of the promising subjects in the investigation of the evolution of the mineralization process.

The "biologically induced" and "biologically matrix mediated" biomineralization types defined by H. Lowenstam (7) have different structural expression. The first type is found in prokariots, "Algae" and evolutionarily may be primary. The mineral secretion is intracellular, extracellular, intercellular. The aggregation of mineral secretions is irregular like non-organic deposition from solutions. The biomineralization of this type of most cases do not fulfil morphogenetic skeletal function. The skeletal biomineralization in some "Algae" (coccolithophorids, diatoms, silicoflagellates) and in animals is of the "matrix mediated" type. Structurally, it is expressed by the presence of a mylti-levelled hierarchical co-subjection of crystalline units, the morphology of which is controlled by organic matrices and can differ morphologically from the inorganic crystals in the presence of structurally different parts and skeletal elements, in a more close integration of a skeleton and a "soft tissue". The beginning of the carbonate "induced" biomineralization, probably, can be connected with the appearance of stromatolites, while the beginning of the "matrix mediated" type is connected with the large-scale development of skeletal faunas in the Early Cambrian. The structural study of ancient organisms, which appeared in the skeletal form independently and in parallel, will help in defining evolutionary ways and laws of the "matrix mediated" biomineralization. The structure of archeocyates, this first wide_spread group of organisms with a compound carbonate skeleton, indicates that the "induced" mineralization could have been the original skeletal type in animals, as well. This is confirmed, in particular, by the test structure of fusulinids, which are referred to the first group of foraminifers with a secretion calcareous skeleton.

Within the "matrix mediated" biomineralization there are two modes of secretion of skeletal material, namely unicellular and tissue ones. The first mode is spicular skeletogenesis in sponges and in some species of coelenterates, as for the advanced invertebrates this mode occurs in the larval skeletogenesis of echinoderms and in the secondary fibrous shell layer in articulate brachiopods. The next mode is a common secretion mode of shell layers in molluscs and other invertebrates. The unicellular secretion mode is considered to be primitive, while the tissue one is advanced.

Some types of skeletal microstructures are considered transitional, for example, a crossed-bladed one. It was characteristic of some fossil groups of the Palaeozoic brachiopods (Davidsoniacea, Strophomenacea) and it was absent in the Mesozoic and Cenozoic forms. On the whole, the tissue secretion mode is not developed in brachiopods. The presence of structures relative to that indicated above was found in some bivalve moluscs (4). This fact, along with the data on unicellular spicular biomineralization of the periostracum of mytilids and of the perinotum of chitons, suggests the unicellular type of primary secretion of the mineral skeleton in molluscs. The study of shell structure of the Early Palaeozoic molluscs and mollusc-like forms will check this hypothesis and contribute to the data on the early evolution and biomineralization regularities of this type. Structural transformations of skeletal parts of organisms, which took place in the evolutionary series of forms expressed in phylogenetic structural trees, no doubt, are to be interpreted in the terms of biomineralization. The geologically recorded successive origins of large taxa of foraminifers, the secretion calcareous shell of which being of different microstructure types, indicates, probably, transition from the "induced" biomineralization to the "matrix mediated" one and the improvement of the latter. Fusulinids are the first representatives with a real secretion calcareous shell, known since the Devonian. Their shell walls have microgranular microstructure, obviously, resembling that of many Palaeozoic Algae and archaeocyathes (Tabl. 1). This structure can indicate biomineralization of the "induced" type. In miliolids, which appeared later, the porcellaneous microstructure of the shell wall, consisting of disoriented bundles, or clusters of calcareous rods, in very similar to that of many recent algae, which is also considered by Lowenstam to be derived from the "induced" type of biomineralization. The wall microstructure of more advanced taxa of foraminifers (the radial-rayed type) known since the Mesozoic, no doubt, is of the "matrix mediated" origin.

The affinity of structural skeletal peculiarities in unrelated groups of organisms, and parallel and independent appearance of some microstructures, obviously, reflect regularities of the biomineralization process and its evolution as an all-biological phenomenon. The common features are such as the universal presence of spherulitic and prismatic microstructures in all invertebrates, the independent acquirement of the nacreous structure by primitive groups of molluscs of different classes, the independent and unsimultaneous origin of the crossed lamellar and homogeneous microstructures among them, the regular reduction of the organic components in the most advanced microstructural types.

The latter trend, which is most clearly seen while comparing "the primitive" nacreous layer with "the advanced" homogeneous one, indicates either a quantitive reduction of the organic fractions in the mineralization process, or the exclusion of organic components, unrelated to biomineralization, from the skeletal composition, or both. In any case, this, obviously, reflects the biomineralization "progress".

Being characteristic of the "matrix mediated" type, the differentiation of microstructures between separate elements of one and the same skeleton (between the wall, septa, siphonal wall in cephalopods, between the shell wall and elements of brachial and dental apparatus in brachiopods, etc.), and differences in structural changes of these elements in phylogenetic series reveal new possibilities of homologization of elements. This new conception of homology (the physiological homology (1)), proved by ideas about cyclic development of the secretory phases in skeletal formation of molluscs (9), and a conception of the regime of mantle secretion activity in brachiopods (15) appear to be useful for the interpretation of phylogenetic relations within separate groups and reflect the variety of biomineralization processes.

The microstructural researches of fossil material face great difficulties caused by secondary diagenetic alterations of primary structures. The problem of postsedimentary alterations of skeletal structures requires special experimental and theoretical investigations. Some regularities of these processes, applied to the nacreous layer in molluscs, were revealed in a series of experiments conducted by Ch. Gregoire (5).

Essential structural changes can take place without the recrystallization and changes of primary mineralogy and composition. On the other hand, fine details of primary structure can be preserved for recrystallization and replacement. Mutvei's data on the structure of the phosphatized nacre of fossil cephalopods is a classical example of it (10). All this proves that, due to their structural peculiarities and the presence of an organic matrix being complex in structure and composition, the postsedimentary alterations of skeletal biominerals differ greatly from the diagenetic changes of non-biologically formed minerals. The term "paleization", proposed by M. Florkin for transitional fossilization changes of the organic skeletal phase, probably, is more correct when applied to the transformational structural characteristics of biominerals, then the term "diagenesis". The idea of the role of solid-phased processes and organic matrices in fossilization requires further development (3).

The next trend of paleontological researches of the problem of mineralization is paleobiochemical studies, or studies of the organic constituent of skeletal remains of organisms by methods used in biochemistry and mollecular biology. The recent data on occurence of fossil organic molecules in skeletal remains, which have preserved their immune activity, the calcium-binding and other characteristics, also the close study of organic components co-acting in the biomineralization process in recent forms, will let achieve a new level of comparative researches of fossil and recent organisms, a new level of homologization, i.e. the homology of molecules co-acting in the biomineralization process.

REFERENCES

1. Barskov I.S. Studies of cephaloped skeleton structure -Bull. Moscow obschcestva ispytatelei prirody, 1973, ser. geol., N 4 (in Russian).
2. Vinogradov A.P. Chemical elemental composition of marine organisms. Moscow. "Nauka", 1, 273 c. (in Russian).

3. Gulubev S.N. Real crystalls in coccolitophorid skeleton. Moscow. "Nauka", p. 200 (in Russian).
4. Carter J.G. Comparative shell microstructure of the Mollusca, Brachiopoda and Bryozoa. -Scann. Electron Microscopy. SEM Inc. AMF O'Hare, 1979, v. 2, p. 439-446.
5. Gregoire Ch. Experimental alteration of the shell by factors involved in diagenesis and in metamorphism. Pt. III. - Thermal and hydrothermal changes in the organic and minteral components of the mural mother-of-pearl. -Bull. Inst. r. Sci. nat. Belg., 1972, p. 85.
6. Hohenegger J., Piller W. Wandstructuren und Grossgleiderung der Foraminiferen.-Sitzungzber. Osterr. Akad. Wiss., Mathem. -naturwiss. Kl., 1975, Abt. I, Bd 184, N 1-5, S. 67-89.
7. Lowenstam H.A. Minerals formed by organisms. -Science, 1981, v. 221, N 448, p. 1126-1131.
8. Lowenstam H.A., Margulis A. Calcium regulation and the appearance of calcareous skeleton in the fossil records. - In "The mechanisms of biomineralization in animals and plants". Proc. 3rd Intern. biomineral. sympos. Tokyo, 1980, p. 299-300.
9. Mutvei H. On the shells of Nautilus and Spirula with notes on the shell secretion in non-cephalopod molluscs. -Arkiv zool. Svenska Vetenskapsakad., 1964, v. 16, N 14.
10. Mutvei H. Ultrastructural evolution of molluscan nacre. -In "Biomineralization and biological metall accumulation". Westbroek J. and de Jong E.M. (eds), Reidel Publ. Comp. Dordrecht, 1983, p. 267-271.
11. Ross C.A., Ross J.R.P. Adaptive evolution in the soritids Marginopora and Amphisorus (Foraminiferida). -Scann. Electron Microscopy. SEM Inc., AMF O'Hare, 1979, v. 2, p. 53-60.
12. Rozanov A.Yu., Sayutina T. Microstructure of certain algae, archaeocyathids and cribricyathids. -3rd North Amer. Paleont. Convent. Proc., 1982, v. 2, p. 453-455.

13. Sorauf J.E. Original structure and composition of Permian rugosa and Triassic scleractinian corals. -Paleontology, 1978, v. 21, p. 321-339.
14. Vacelet J. Description et affinites eponge sphinctozoaire Actuelle. -In "Biologie des spongiaires". Colloques internat. CNRS, 1979, 291, p. 483-493.
15. Williams A.A. A history of skeletal secretion among articulate Brachiopods. -Lethaia, 1968, v. 1, N 3, p. 268-287.

FIGURE 1. (1) Fossil algae _Renalcis novus_; Early Cambrian; Mongolia, Salaygol (from Rozanov, Sayutina, 1982 (12)); (2) Archaeocyate _Dictyicyathus salairicus_; Early Cambrian; USSR, river Lena (from Rozanov, Sayutina, 1982 (12)); (3) Sphinctozoan sponge _Verticellites convexus_; Callovian, Jurassic; USSR, Pamir (foto A. Zhuravlev); (4) Test wall of foram _Rugofusulina alpina communis_; Later Carboniferous (from Hohenegger, Piller, 1975 (6)); (5) Sphinctozoan sponge _Vaceletia cripta_; Rec.; New Kaledonia (from Vacelet, 1979 (14)); (6) Algae _Galazaura obtusata_; Rec. from Lowenstam, 1981; (7) Test wall of foram _Marginopora vertebralis_; Rec.; Australia, Great Barrier Reef (from Ross, Ross, 1979 (11)). Note: The microstructures of fossil algae and "primitive" animals are similarly constructed by randomly orientired grains. The recent algae, sphinctozoan sponge and test wall foram (order Miliolida) microstructures are similar so. Its have random disposition of ajacicular rods. The grained structure may be diagenetic altered rod structure.

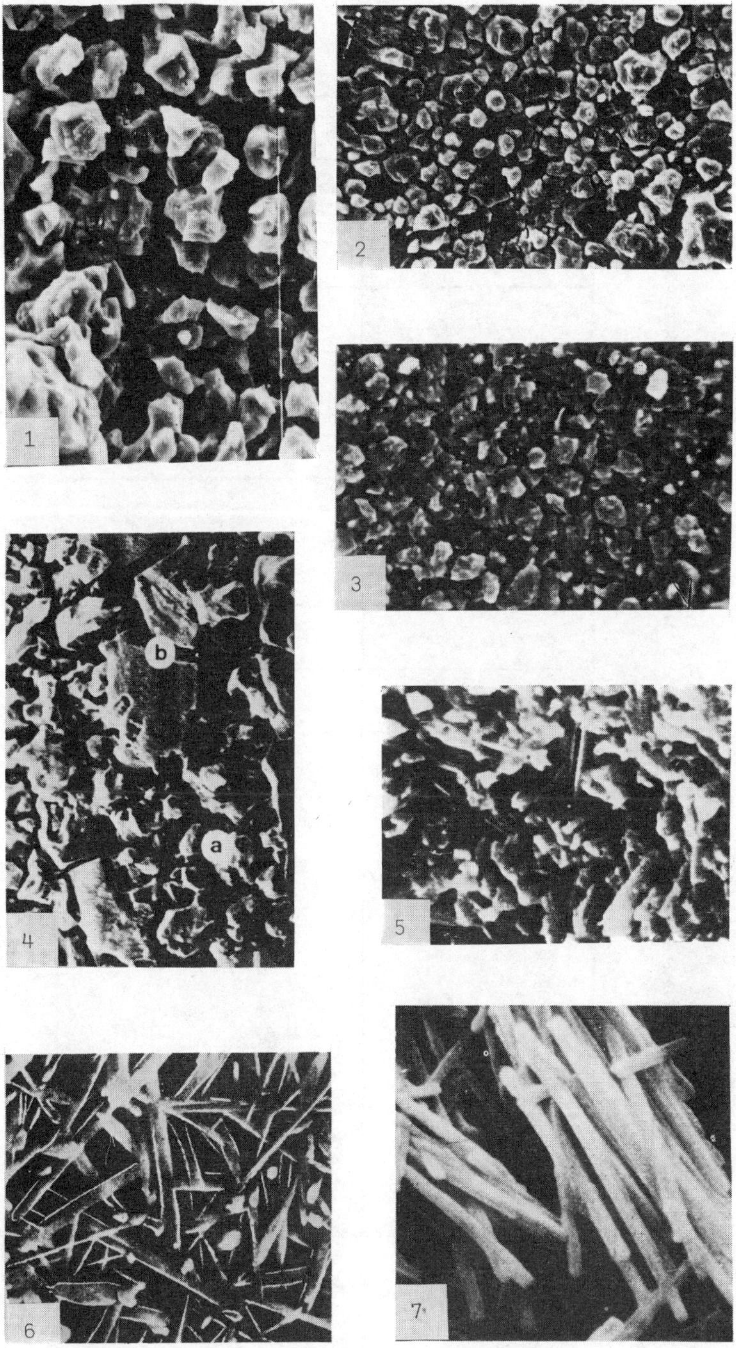

TABLE 1. Distribution of the skeletal biomineralisation types in main phyla.

Type of skeletal biomineralisation \ Taxons	"Algae"	Radiolaria	Foraminifera	Coelenterata	Porifera	Annelida	Mollusca	Arthropoda	Bryozoa	Echinodermata	Brachiopoda	Chordata
Phosphate								?+			+	+
Carbonate	+		+	+	+	+	+	+	+	+	+	
Siliceous	+	+			+							
Others		+										

THE PATTERN OF PHANEROZOIC COMMUNITY EVOLUTION

A. J. BOUCOT
Departments of Geology and Zoology, Oregon State University, Corvallis, Oregon 97331

ABSTRACT
The division of the marine, level bottom fossil record into twelve major, Ecologic-Evolutionary Units is emphasized. Within each of these Units a number of community groups co-occur. Non-level bottom community types, such as parts of the reef community complex, have stratigraphic ranges that are commonly less than those of the ecologic-evolutionary units within which they occur. Higher trophic level marine units may be decoupled from low trophic level units. The ecologic-evolutionary units of the terrestrial environments are clearly decoupled from those of the marine environments, although not yet well defined nearly as well as those for the marine environments.

I provided (1983a) a summary concerning the overall pattern of Phanerozoic Community Evolution. This contribution is partly a repetition and rephrasing of that summary, together with some additional data of concern.

Figure 1 (taken from Boucot, 1983b) is a diagram indicating the overall pattern of community group (the term "community group" is equivalent to the term "biofacies" _if_ narrowly construed, i.e., _Stringocephalus_ biofacies, _but not_ shelly facies) behavior. A community group consists of the data of community evolution, within any specific community "type", arrayed on a time ordinate. Figure 1 also shows the relative abundance (by means of line widths; wider lines represent numerically more abundant genera and their species) of the co-occurring genera and their species.

In 1983a I tried to make the point that the fossil record may be profitably considered in terms of the stratigraphic ranges of co-occurring community groups. I emphasized that there is an overwhelming tendency for

Proceedings of the 27th International
Geological Congress, Volume 2, pp. 13–21
PALAEONTOLOGY

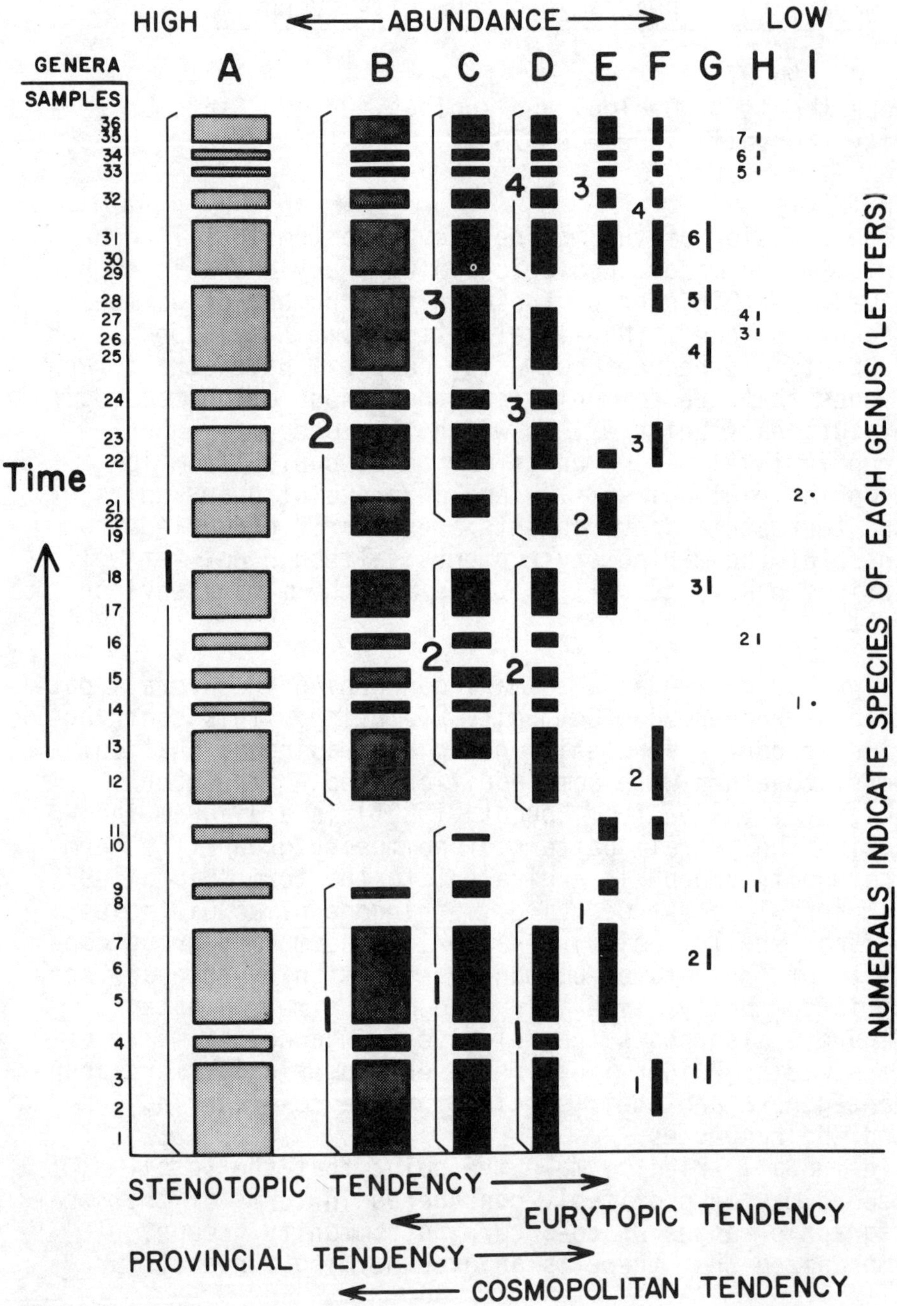

HIGH
ABUNDANCE
LOW
GENERA
SAMPLES
A B C D E F G H I
Time
NUMERALS INDICATE SPECIES OF EACH GENUS (LETTERS)
STENOTOPIC TENDENCY
EURYTOPIC TENDENCY
PROVINCIAL TENDENCY
COSMOPOLITAN TENDENCY

FIGURE 1. Diagram (taken from Boucot 1983b) indicating the characteristics of a community group through an interval of geologic time. Sampling gaps (indicated on figure) may occur because of nondeposition, erosion, collecting failure (commonly caused by covered intervals), or presence of another community group because of local or regional environmental shifts. Line widths proportional to specimen abundances. Letters symbolize genera. Numerals symbolize species of genera in each lettered column, i.e., species replacement. Specimen abundances tend to be fairly constant, as diagrammed, over time; rare taxa tend to always be rare through time, abundant taxa tend to be abundant through time, and so on. The shading used on the lines is for artistic purpose only. Note that the species of the more abundant, common genera evolve slowly, and that the reverse is true for the species of the less common genera. The time interval indicated is of the order of magnitude of 5 to 10, or 20, millions of years for the average level-bottom, marine benthic community group.

level bottom community groups to have common upper and lower stratigraphic limits. All I did here was to rephrase the common knowledge of biostratigraphers that many level bottom biofacies have: 1) definite, known stratigraphic ranges; 2) that the definite, known stratigraphic ranges of most level bottom biofacies (community groups in my terminology) tend to be similar to those of many others with which they co-occur. In actual practice this means that there are about twelve major, Phanerozoic community group intervals (I term them Ecologic-Evolutionary Units) as shown in Figure 2. These intervals (Boucot, 1983a), characterized by a large number of co-occurring biofacies, are familiar to the biostratigrapher.

I also (1983a) emphasized that varied, non-level bottom biofacies, such as the reef complex of communities, varied thicket types (crinoidal, bryozoan, etc.), and forest types (sponges, etc.), and the like occur within the same stratigraphic limits as these twelve major Ecologic-Evolutionary Units. But, the non-level bottom types almost always have shorter stratigraphic ranges than those possessed by the level bottom community groups. These shorter ranges generally involve a later initiation of the non-level bottom community groups, and commonly involve the same termination. The terminations are commonly, but not always, coincident with major extinction events.

I emphasized that the major changes in marine level bottom Ecologic-Evolutionary Units commonly do not coincide with those seen on land. As an example of this "decoupling" of the marine biofacies from the nonmarine I used the well known fact that the stratigraphic ranges of the higher land plant based time units Palaeophytic, Mesophytic and Cenophytic do not coincide in any way with the marine Ecologic-Evolutionary Units, nor with the terms Paleozoic, Mesozoic, and Cenozoic. It is unfortunate that at this time we have no comprehensive synthesis in community group terms for the nonmarine biota. Therefore, comparisons between the marine and nonmarine for contemporaneity of major changes is difficult to make. However, it is clear that the major change in the tetrapods between the "Paleozoic" and "Mesozoic" types takes

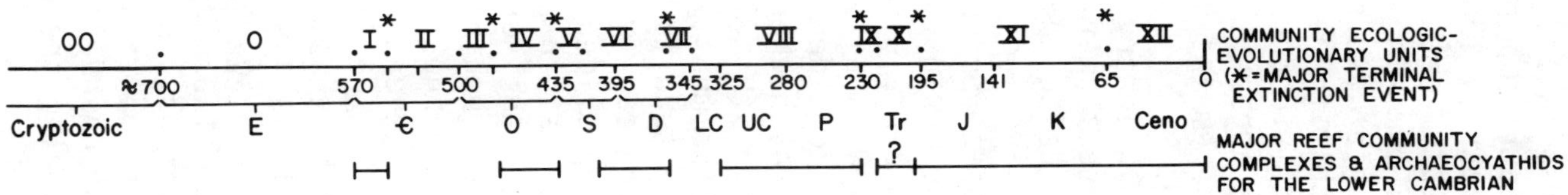

FIGURE 2. Diagram indicating the stratigraphic ranges of the Phanerozoic Ecologic-Evolutionary Units, with additions for the Ediacaran (unit "0"), and Cryptozoic (unit "00"), and of the major reef community complexes and archaeocyathid community complexes (taken from Boucot, 1983a, Fig. 3). Major extinction events in the level-bottom marine world are indicated with asterisks.

place within the Middle and Late Triassic, i.e., the extinction of most Paleozoic amphibian and reptilian types and their replacement by archosaurian types with a Mesozoic flavor (Charig, 1980; Bonaparte, 1983). Following on these Paleozoic-Mesozoic events comes the extraordinary radiation of the placental mammals during the Cenozoic. Thus, despite our ignorance of overall terrestrial community groups, it is clear that terrestrial ecologic-evolutionary units are essentially decoupled from those of the marine environment. The apparent decoupling of woody plant from tetrapod units remains currently unexplained.

The twelve, major ecologic-evolutionary units of the Phanerozoic pertain, as mentioned to marine, level-bottom benthos. Most of the organisms being considered here are megafossils--not enough data has yet accumulated concerning the ecologic-evolutionary patterns shown by many microfossil groups. Most of the marine, level-bottom, megafossils being considered here are also low trophic level organisms--they include few carnivores.

I am not very well prepared to consider the ecologic-evolutionary unit pattern shown by higher trophic level, marine carnivores. The reason for this is that little attention has been shown to the community ecology of such organisms, probably because they tend to be much less abundant as specimens than is the case with low trophic level organisms. However, data provided in Harland et al. (1967) for the fishes indicates that in at least one instance, a clearly higher trophic level group was decoupled from the low trophic level benthic invertebrates. In particular the placoderms and dipnoans of the Devonian show a marked terminal extinction event at the end of the Devonian rather than at the end of the Frasnian as is the case with the low trophic level benthos. The reasons for this distinction are unknown. It is also interesting to note that data provided by the same source indicates a major adaptive radiation of the actinopterygians at the beginning of the Lower Carboniferous, as is also the case with the lower trophic level benthos, and another for the teleosts and many of the chondrichthys near the beginning of the Jurassic as is also the case with the lower trophic level benthos. I earlier (1975) pointed out that the marine eurypterids became extinct near the end of the Silurian, after which time they were restricted to the nonmarine environment for the remainder of the Paleozoic. This terminal extinction of a major marine group does coincide with a relatively minor terminal extinction of certain low trophic level marine benthos, but the disparity here is marked.

The fact that many animal and plant groups show evidence, both today and from the fossil record, for gradual size increase during evolutionary time, could be construed as evidence against the constancy of my community group concept. That is, if certain taxa evolve towards larger (or smaller) size it is only reasonable to conclude that their abundance relative to co-occurring taxa within the same community group should change correspondingly. We are all aware, for example, that large species within any group generally have far lower population densities than small species. Large clams are forced by pure geometry to have a lower population density than small clams, and so forth. Earlier (Boucot, 1976) I provided some comments on data from the low trophic level, marine, fossil megabenthos. There are specific intervals, such as the later Ordovician, Early Devonian and Late Cretaceous, when many, unrelated groups of such marine benthos show an overall tendency towards gigantism following Cope's Rule. During such time intervals if most taxa show an increase in size there would be an overall lowering of numbers of individuals of most taxa per area, but the relative numbers of each taxon could stay about the same. However, there are certainly examples from the fossil record where this is not the case. For example, there is a marked trend during the Silurian for pentameroid brachiopods in several lineages (*Stricklandia-Costistricklandia*; Pentamerinidae; some gypidulids) for great, evolutionary increase in size. But, these taxa commonly make up almost monotypic, very high dominance, roughwater communities. Therefore, although there will be marked lowering in numbers of individuals per area, the overall composition of the community group will not change, i.e., the same taxon will remain dominant even though its overall numbers go down. But, there are also examples in which a high diversity community group will be the site for a Cope's Rule increase in size of one of the taxa. In such a case we face an apparent exception to my community group, community evolution generalization. But, even here if we pay attention to both the biomass of each taxon within the evolving community group, as well as overall productivity rather than pure numbers of individuals it may be that the generalization still holds. We cannot, of course, measure productivity in the fossil record, although we can measure shelly biomass. But, if one assumes that relative rates of productivity in low trophic level seashells of the past were similar to similarly sized shells today, we can make some estimates--particularly if we stay within the same environments and similar climatic zones. The results of such extrapolation is to conclude that things stay about the same.

Another example of decoupling has been pointed out by Douglas and Woodruff (1981) who report that benthic abyssal foraminifera do not show a marked effect of the terminal Cretaceous extinction event as is commonly the case in shallower waters. The reasons for this decoupling are uncertain.

All of the above does indicate that there is enough evidence for decoupling between major portions of the global ecosystem to suggest that the factors responsible for both major terminal extinctions and adaptive radiations have not been global in all environments. This fact would tend to further strengthen the position that extraterrestrial factors have probably not been important here during the Phanerozoic.

REFERENCES

Bonaparte, J. F. (1982). Faunal replacement in the Triassic of South America. J. Vert. Paleo., 2, 362-371.

Boucot, A. J. (1975). Evolution and Extinction Rate Controls. Elsevier, 427 pp.

Boucot, A. J. (1976). Rates of size increase and of phyletic evolution. Nature, 261, 694-696.

Boucot, A. J. (1983a). Does Evolution Take Place in an Ecological Vacuum? II. J. Paleo., 57, 1-30.

Boucot, A. J. (1983b). Area-Dependent-Richness Hypotheses and Rates of Parasite/Pest Evolution. Am. Nat. 121, 294-300.

Charig, A. J. (1980). Differentiation of lineages among Mesozoic Tetrapods. Mem. Soc. Geol. France, N. S., T. LIX, Mem. 139, 207-210.

Douglas, R., and Woodruff, F. (1981). Deep-sea benthic foraminiferida. In: C. Emiliani, ed., The Sea, vol. 7, The Oceanic Lithosphere, John Wiley, 1233-1327.

Harland, W. B. et al. (1967). The Fossil Record. Geol. Soc. London, 827 pp.

EVOLUTIONARY, GEOGRAPHIC, AND ECOLOGICAL ASPECTS OF THE RISE OF ANGIOSPERMS

JAMES A. DOYLE
Department of Botany, University of California
Davis, CA 95616, USA

ABSTRACT
Fossil pollen and leaf records document progressive diversification of angiosperms during the Early Cretaceous, beginning with monosulcate pollen and simple leaves, with the tropical belt (Africa-South America) serving as the most important center of origin and diversification of new types. Pre-angiospermous tropical plant communities were relatively species-poor; factors contributing to the angiosperms' exceptionally rapid diversification and replacement of earlier groups in the tropics may include insect pollination, allowing more dispersed distributions of individuals and escape from herbivores and pathogens, superior colonizing abilities, vessels, and higher intrinsic speciation rates.

Paleobotanical and palynological studies of the past 25 years have greatly clarified the timing and course of early angiosperm evolution and have provided evidence on the geographic spread and ecology of early angiosperms. New data contradict earlier ideas that angiosperms were already highly diversified when they first entered the Early Cretaceous fossil record, although they do leave open the possibility that very rare, primitive angiosperms existed before the Cretaceous (Doyle, 1969, 1978; Muller, 1970). Comparisons with geographic trends in the rest of the flora suggest explanations for the rapid displacement of the older gymnosperm- and fern-dominated vegetation by angiosperms, one of the most radical transformations in the history of life.

EVOLUTIONARY EVENTS AND GEOGRAPHIC PATTERNS

The undisputed record of angiosperms begins in the Barremian Stage, with the appearance of rare monosulcate

Proceedings of the 27th International
Geological Congress, Volume 2, pp. 23–33
PALAEONTOLOGY

pollen grains which differ from the monosulcate pollen of many gymnosperms in their reticulate sculpture, columellar infratectal structure, and lack of a laminated endexine (Doyle et al., 1975). Among modern angiosperms,

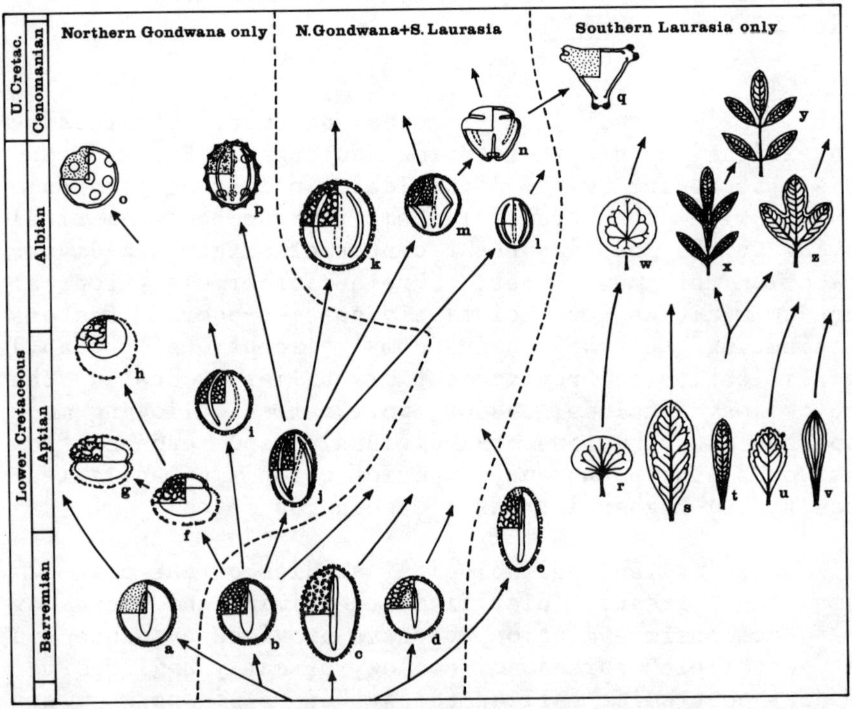

FIGURE 1. Stratigraphic and geographic distribution and relationships of major Early Cretaceous angiosperm pollen and leaf groups. a: tectate-granular relatives of Clavatipollenites; b: Clavatipollenites and Retimonocolpites; c: Stellatopollis; d, e: Liliacidites spp.; f: Afropollis operculatus; g: A. zonatus; h: A. jardinus; i: monosulcate with extended sulcus; j: early tricolpate; k: large, sculptured tricolpate; l: small, smooth tricolpate; m: tricolporoidate; n: oblate-triangular tricolporate; o: polyforate; p: tricolpodiorate; q: early Normapolles; r: Proteaephyllum reniforme; s: Ficophyllum; t: Rogersia; u: serrate leaf; v: Acaciaephyllum; w: Nelumbites; x: pinnately dissected Sapindopsis; y: compound Sapindopsis; z: platanoid.

monosulcate pollen is restricted to monocots and dicots of the subclass Magnoliidae and has long been considered primitive by comparative morphologists (Walker, 1976). Barremian and Aptian angiosperms of Europe and North America show considerable diversity in exine structure, but all at a monosulcate level (Kemp, 1968; Doyle et al., 1975; Hughes et al., 1979): finely to coarsely reticulate (including *Clavatipollenites* Couper, the first type recognized: Fig. 1b), with triangular supratectal projections (*Stellatopollis*: Fig. 1c), and with finer sculpture at the poles or the ends of the grain (*Liliacidites*: Figs. 1d,e), a trend characteristic of monocots rather than magnoliid dicots today (Doyle, 1973). Rare reticulate-columellar monosulcates from the Upper Triassic (Cornet, 1977) may represent still older angiosperms, but the fact that they have a thick, apparently laminated endexine and occasional accessory furrows, as in the Mesozoic gymnosperm genus *Eucommiidites*, suggests they may instead represent a convergent, extinct gymnosperm group (Doyle, 1978).

Barremian angiospermous monosulcates are even more abundant and diverse in the Early Cretaceous tropical belt (Africa-South America, the Northern Gondwana province of Brenner, 1976; Doyle et al., 1977). Although *Clavatipollenites* in the strict sense and *Liliacidites* species with fine ends are not definitely recognized here until later, these floras do contain other reticulate monosulcates, *Stellatopollis*, *Liliacidites* with fine poles, and, most interesting, a variety of forms related to *Clavatipollenites* on shape, thickness of the nexine, and aperture sculpture, but with a continuous tectum and apparently granular infratectal structure (Fig. 1a), characters interpreted as more primitive than those of *Clavatipollenites* by students of modern Magnoliidae (Walker, 1976; Le Thomas, 1981).

Angiosperm leaves from the monosulcate phase of the record are best known from Aptian(?) beds of the lower Potomac Group of the eastern United States (Hickey and Doyle, 1977). These rare leaves vary in shape from reniform to elliptical to narrowly obovate (Figs. 1r-t), and from entire-margined to serrate (Fig. 1u) and lobulate, but they are all apparently simple and basically pinnately veined, and their venation shows unusual irregularity in course and differentiation into orders, as in some modern Magnoliidae. Some have long bases and apically fused venation, suggestive of monocot affinities (Doyle, 1973; Fig. 1v).

With a few exceptions (e.g., Vakhrameev and Kotova, 1977), angiosperms are almost unknown from the Barremian-Aptian of higher latitude areas.

The second major event in the angiosperm pollen record is the appearance of tricolpate pollen, the type basic to all "higher" dicot subclasses (i.e., except Magnoliidae). The first tricolpates appear near the base of the Aptian in Northern Gondwana (Brenner, 1976; Doyle et al., 1977). Their reticulate-columellar exine structure supports derivation from older angiospermous monosulcates, and they show irregularities in orientation of the three furrows which suggest an origin by extension, spiralization, and fragmentation of a single sulcus (Fig. 1j). Other new pollen complexes which deviate from the monosulcate condition appear at about the same time, such as <u>Afropollis</u> (Doyle et al., 1982), older species with a ring-furrow at one pole or at the equator (Figs. 1f,g), later species with no aperture (Fig. 1h), and grains with the sulcus extended lengthwise around the grain (Fig. 1i). Except for a few isolated occurrences, none of these groups are known outside Northern Gondwana until the early Albian, when tricolpates begin to appear consistently in Southern Laurasia (Kemp, 1968; Hickey & Doyle, 1977). Tricolpates enter Australia in the middle Albian, just after <u>Clavatipollenites</u> (Burger, 1980), and Canada still later, in the late Albian or even the Cenomanian (Brenner, 1976).

Tricolpate angiosperms diversify and increase in abundance in the Albian of both Gondwana and Laurasia, leading by the end of the stage to large, small, smooth, and coarsely sculptured tricolpates (Figs. 1k,l), tricolpates with rudimentary ora (tricolporoidates: Fig. 1m), and oblate-triangular tricolporates (Fig. 1n). Again, there is a greater variety of new types in Northern Gondwana, such as polyforates (Fig. 1o) and tricolpodiorates (Fig. 1p). Middle Cenomanian beds of Laurasia yield the first members of the triporate Normapolles group (Fig. 1q), apparently derived from triangular tricolporates. The Normapolles are believed to include the ancestors of most modern Amentiferae (Juglandaceae, Betulaceae, etc.), groups considered advanced under orthodox theories of angiosperm evolution (Doyle, 1969, 1978; Muller, 1970).

Angiosperm leaves from the phase of tricolpate expansion are known from a great number of localities, mostly in Southern Laurasia (Potomac Group, Kansas, Portugal, Kazakhstan, Soviet Far East), and they include a variety of new morphological complexes (reviewed by Hickey and Doyle, 1977). Among these are cordate and peltate leaves (Fig. 1w), some apparently related to modern Nymphaeales (Magnoliidae); pinnately dissected and later pinnately compound leaves (Figs. 1x,y), probably early members of the basically tricolpate subclass Rosidae; and palmately lobed leaves with venation suggestive of relationships with modern tricolpate Platanaceae (Fig. 1z). These leaves show a marked trend to more regular venation, with a well-defined hierarchy of vein orders. Studies on the Dakota Group of Kansas (Dilcher, 1979) show that angiosperms had also attained considerable diversity in reproductive structures by the early Cenomanian, including magnolioid floral axes with numerous free follicles, platanoid inflorescences made up of small, inconspicuous flowers, and even syncarps.

ECOLOGICAL INFERENCES

Present palynological data favor the concept that angiosperms originated and diversified most vigorously in the tropics and spread into higher latitude areas, as first argued by Axelrod (1959) on megafossil data, mostly from within Laurasia. Brenner (1976) drew similar conclusions from the observed poleward spread of tricolpates, but this inference applies directly only to one subgroup of angiosperms, the higher dicots. However, the greater diversity of angiospermous monosulcates in the Barremian of Northern Gondwana, including possibly more primitive types, does suggest that this is the most likely area for origin of the angiosperms as a whole.

The sedimentary facies distribution of early angiosperm

leaves in Southern Laurasia has been interpreted as evidence for Stebbins's (1974) hypothesis that the first angiosperms were fast-growing, weedy shrubs which originated under selective pressures for rapid reproduction in unstable habitats of seasonally arid tropical regions, rather than trees of wet tropical forests (Hickey and Doyle, 1977). In the lower Potomac Group, angiosperm leaves are usually found in relatively coarse-grained facies presumably deposited near stream margins, rather than in fine-grained, organic-rich flood plain and swamp facies, which contain abundant fern and gymnosperm remains. By the end of the Albian, angiosperm leaves have become locally dominant and occur in a wider range of facies (e.g., peltate aquatics in presumed pond deposits), but they are still typically associated with coarser sediments or with facies shifts reflecting local disturbance, and the continued subordinate role of angiosperm pollen suggests that gymnosperms still dominated the regional vegetation (Hickey and Doyle, 1977; Retallack and Dilcher, 1981). Although Southern Laurasia was presumably outside the main center of angiosperm evolution and relatively wet in climate, these data do indirectly support Stebbins's hypothesis, since stream margins are disturbed habitats to which weedy immigrants from dry tropical areas might be preadapted. The early occurrence of presumably herbaceous monocots and aquatics is also easier to explain in terms of a weedy ancestry (Hickey and Doyle, 1977).

Given the indications that angiosperms originated in Northern Gondwana, data from this region should provide more direct evidence on their original ecological preferences. Paleogeographic reconstructions indicate that Northern Gondwana lay in the tropical zone, and several authors have argued that its climate was semiarid rather than wet (Brenner, 1976; Doyle et al., 1977). Arguments for aridity include the occurrence of Aptian salt deposits on both sides of the South Atlantic (Brazil, Gabon, Angola) and the curiously depauperate character of the non-angiospermous flora compared with that of Southern Laurasia: the paucity of ferns, except for the xeromorphic genus Weichselia; the dominance of xeromorphic cheirolepidiaceous conifers (Classopollis), in contrast to the Pinaceae, Taxodiaceae, Podocarpaceae, and ginkgophytes abundant at higher latitudes; and the abundance of polyplicate ephedroid pollen, believed to include ancestors of the modern xerophytic gymnosperms *Ephedra* and *Welwitschia* (Gnetales). Closer examination, however, suggests that Northern Gondwana included not only dry areas (e.g., the salt basins), but also wetter ones, apparently corresponding to the equatorial rain forest belt (Doyle et al., 1982). This belt contains coal deposits, no salts, and more abundant ferns and Araucariaceae, but *Weichselia* and Cheirolepidiaceae are still abundant. Angiosperms are common throughout the region, with some groups (e.g., *Afropollis*) more abundant in the presumed wet belt.

These observations suggest that the low diversity of gymnosperms and ferns in Northern Gondwana was due not to aridity but rather to intrinsic limitations of their biology under tropical conditions, while the fact that angiosperms were more diverse from the start in the tropics, as they are today, may be a result of evolutionary innovations which allowed them to overcome these limitations (Doyle et al., 1982). One such innovation is insect pollination, which the sculpture of early angiosperm pollen suggests was already established by the Barremian. Gymnosperms, being wind pollinated, are constrained to form relatively dense stands in order to maintain outcrossing, limiting the number of species in a given area and subjecting them to increased levels of herbivore and pathogen attack, which is especially severe in non-seasonal tropical habitats. Conversely, by making possible more scattered distributions of individuals, insect pollination would allow angiosperms to escape herbivores and pathogens (and hence to expend less on chemical defenses) and to pack more species into a given area (Janzen, 1970; Regal, 1977). In general, the more rapid, efficient reproductive cycle in angiosperms would allow them to exploit ephemeral, disturbed habitats in both tropical and temperate zones more effectively than gymnosperms and ferns. Another angiosperm innovation of special value in the tropics is the presence of vessels in the xylem, which, by allowing more rapid conduction of water under conditions of high evaporative demand, would allow angiosperms to produce larger, more efficient leaves than gymnosperms and ferns, which had to compensate for the inefficiency of their vesselless xylem by being xeromorphic. In addition, several features of angiosperms may have given them intrinsically higher speciation rates than other groups (cf. Gould and Eldredge, 1977), such as the potential for new isolating mechanisms based on pollinator behavior and stigmatic blocks to pollen tube growth, and hence accelerated their rise to the high levels of species diversity made possible by their other features. Any or all of these factors may have contributed to the rapid rise of angiosperms and their ultimate displacement of older groups, especially in the tropics. Because of lower predation pressures and evaporative demand under temperate conditions, gymnosperms may have been at less of a disadvantage at higher latitudes, where they often compete successfully with angiosperms to the present day.

Analogies with the history of another group, the producers of ephedroid pollen (presumably including ancestors of the modern Gnetales, _Ephedra_, _Welwitschia_, and _Gnetum_), may be useful in evaluating this scenario. Ephedroids are known since the Triassic, but their main expansion occurs in the Early Cretaceous in the same area (Northern Gondwana) where angiosperms are most abundant (Doyle et al., 1982). Judging from the morphology of their modern representatives, Gnetales are not related to angiosperms--their compound strobili, bilateral seeds, and anatomy indicate relationships with coniferopsid gymnosperms, whereas angiosperms appear to be cycadopsid (seed fern) derivatives (Doyle, 1978)--but they show parallel trends to angiosperms in possession of vessels, acceleration of the life cycle, and, in _Gnetum_, remarkably dicot-like leaves. Perhaps these parallel advances allowed Cretaceous Gnetales to overcome some of the ecological disadvantages of other gymnosperms in the tropics, but they were unable to achieve high levels of species diversity because they remained wind pollinated and had fewer sorts of potential isolating mechanisms. It is suggestive that _Gnetum_ is one of the few modern gymnosperms found in tropical rain forests, but it consists of only a small number of relatively stereotyped species.

REFERENCES

Axelrod, D.I. (1959). Poleward migration of early angiosperm flora. Science 130, 203-207.

Brenner, G.J. (1976). Middle Cretaceous floral provinces and early migrations of angiosperms. In: Origin and Early Evolution of Angiosperms, C.B. Beck (ed). Columbia, New York, pp. 23-47.

Burger, D. (1980). Palynological studies in the Lower Cretaceous of the Surat Basin, Australia. Bur. Min. Res. Bull. 189, 1-106.

Cornet, B. (1977). Angiosperm-like pollen with tectate-columellate wall structure from the Upper Triassic (and Jurassic) of the Newark Supergroup, U.S.A. Am. Assoc. Strat. Palynol. 10th Ann. Mtg. Abstracts, pp. 8-9.

Dilcher, D.L. (1979). Early angiosperm reproduction: An introductory report. Rev. Palaeobot. Palynol. 27, 291-328.

Doyle, J.A. (1969). Cretaceous angiosperm pollen of the Atlantic Coastal Plain and its evolutionary significance. J. Arnold Arboretum 50, 1-35.

Doyle, J.A. (1973). Fossil evidence on early evolution of the monocotyledons. Quart. Rev. Biol. 48, 399-413.

Doyle, J.A. (1978). Origin of angiosperms. Ann. Rev. Ecol. Syst. 9, 365-392.

Doyle, J.A., Biens, P., Doerenkamp, A., and Jardine, S. (1977). Angiosperm pollen from the pre-Albian Lower Cretaceous of Equatorial Africa. Bull. Centres Rech. Explor.-Prod. Elf-Aquitaine 1, 451-473.

Doyle, J.A., Jardine, S., and Doerenkamp, A. (1982). *Afropollis*, a new genus of early angiosperm pollen, with notes on the Cretaceous palynostratigraphy and paleoenvironments of Northern Gondwana. Bull. Centres Rech. Explor.-Prod. Elf-Aquitaine 6, 39-117.

Doyle, J.A., Van Campo, M., and Lugardon, B. (1975). Observations on exine structure of Eucommiidites and Lower Cretaceous angiosperm pollen. Pollen Spores 17, 429-486.

Gould, S.J., and Eldredge, N. (1977). Punctuated equilibria: the tempo and mode of evolution reconsidered. Paleobiology 3, 115-151.

Hickey, L.J., and Doyle, J.A. (1977). Early Cretaceous fossil evidence for angiosperm evolution. Bot. Rev. 43, 2-104.

Hughes, N.F., Drewry, G.E., and Laing, J.F. (1979). Barremian earliest angiosperm pollen. Palaeontology 22, 513-535.

Janzen, D.H. (1970). Herbivores and the number of tree species in tropical forests. Am. Nat. 104, 501-528.

Kemp, E.M. (1968). Probable angiosperm pollen from British Barremian to Albian strata. Palaeontology 11, 421-434.

Le Thomas, A. (1981). Ultrastructural characters of the pollen grains of African Annonaceae and their significance for the phylogeny of primitive angiosperms. Pollen Spores 22, 267-342; 23, 5-36.

Muller, J. (1970). Palynological evidence on early differentiation of angiosperms. Biol. Rev. Cambridge Philos. Soc. 45, 417-450.

Regal, P.J. (1977). Ecology and evolution of flowering plant dominance. Science 196, 622-629.

Retallack, G., Dilcher, D.L. (1981). A coastal hypothesis for the dispersal and rise to dominance of flowering plants. In: Paleobotany, Paleoecology and

Evolution, vol. 2, K.J. Niklas (ed). Praeger, New York, pp. 27-77.

Stebbins, G.L. (1974). Flowering Plants: Evolution above the Species Level. Harvard, Cambridge, Mass., 399 pp.

Vakhrameev, V.A., and Kotova, I.Z. (1977). Drevnie pokrytosemennye i soputstvuyushchie im rasteniya iz nizhnemelovykh otlozheniy Zabaykal'ya. Paleontol. Zh. 1977, 101-109.

Walker, J.W. (1976). Evolutionary significance of the exine in the pollen of primitive angiosperms. In: The Evolutionary Significance of the Exine, I.K. Ferguson and J. Muller (eds). Academic Press, London, pp. 251-308.

FISH-TETRAPOD RELATIONSHIPS

B.G. GARDINER
Queen Elizabeth College, London, W8.

ABSTRACT

The dipnoans are the closest relatives of tetrapods. The Osteolepiformes are a paraphyletic or grade group and the characters shared by *Eusthenopteron* and tetrapods are either convergent or primitive.

RELATIONSHIPS

Most modern textbooks of vertebrate history suggest that the ancestors of tetrapods can be found within the rhipidistians. However, Rosen et al. (1981) using the cladistic method of analysis concluded that the lungfishes and not any of the rhipidistia are the sister-group of the tetrapods.

In this century a variety of characters have been used to relate osteolepiform rhipidistians and tetrapods. The most commonly quoted resemblances are listed in Table 1 and I will confine my remarks to the first six of these since the remainder also occur in other groups of fishes and are not therefore regarded as synapomorphies of rhipidistians and tetrapods.

Dollo (1896:109) suggested that if osteolepid rhipidistians were ancestral to lungfishes and tetrapods they would have a choana, and in 1912 Watson reported the presence of a choana in *Megalichthys*. Today most workers agree with Watson that rhipidistians possess choanae whereas they consider the posterior opening from the lungfish nasal capsule not to be a true choana (Allis, 1932; Jarvik, 1942).

Proceedings of the 27th International
Geological Congress, Volume 2, pp. 35–44
PALAEONTOLOGY

TABLE 1. This table lists the most often cited similarities between rhipidistians and tetrapods used to support the view that tetrapods arose from rhipidistians. The first six are discussed in the text.

1. Choana present.
2. Paired fins of Eusthenopteron most like those of tetrapods.
3. Similar pattern of skull roofing bones.
4. Palate construction and jaw suspension similar.
5. Vertebrae of osteolepiforms similar to those of early tetrapods.
6. Eusthenopteron has double-headed dorsal ribs like tetrapods.
7. Scapulocoracoid element of shoulder girdle like that of early tetrapods with a complex glenoid face.
8. Double-headed hyomandibular, like tetrapod stapes.
9. Humerus of Eusthenopteron with process for muscle attachment.
10. Intracranial joint immobilized in some rhipidistians (intracranial joint lies behind the eye and divides braincase into two portions capable of moving one on the other as in the living coelacanth).
11. Notochord penetrates the base of the braincase.
12. Opercular bones present in primitive tetrapods.
13. Ichthyostega has a fish-like tail.
14. Osteolepiforms have folded teeth of a similar type to those seen in labyrinthodonts.

The key to our new scheme (Rosen et al., 1981) lies in the reinterpretation of the dipnoan internal nostril. Recent lungfishes do not have marginal jaw bones or teeth, thus the morphological relationships of the choana is not immediately comparable. Nevertheless in the Devonian lungfish Griphognathus the internal nostril opened through the bony palate internal to an arcade of maxillary and premaxillary teeth while the anterior opening interrupts the tooth row (between maxillary and premaxillary). These conditions are directly comparable to those in Ichthyostega and to nostril development in Recent urodeles. Thus we postulate that the internal opening in dipnoans is a true choana.

At the same time we re-examined the anatomical criteria for the presumed "choana" in rhipidistians. Although choanae have been restored in many genera, the only unchallenged information that has been presented were Watson's (1926) restoration of Eusthenopteron and his account of Megalichthys and Jarvik's (1942, 1972) more detailed accounts of snout structure in Porolepis and Eusthenopteron. In Porolepis, no palate was described, but in the related Glyptolepis the restored fenestra exochoanalis is very small and in our estimation no more than a notch in the margin of the palate at the junction of four dermal bones. The strongest case for the existence of a choana in rhipidistians comes from Jarvik's (1942) description of the fenestra endo-and exochonalis and the structure surrounding them in Eusthenopteron. Here there is an opening in the bony palate (just below a large fenestra in the ventrolateral wall of the nasal capsule) between the palatine and vomer medially and the premaxillary and maxillary laterally. A similar palatal fenestra is found in Polypterus (a fish without a choana). Moreoever like Polypterus the palatoquadrate in our acid prepared specimen of Eusthenopteron appears to have been movably articulated with the snout. There is no living vertebrate with a joint through the choana. Several osteolepiform genera possess fangs at the tip of the dentary and on the coronoid. In restorations these coronoid fangs appear to fit into the "choana". We therefore conclude that the existence of a choana in Eusthenopteron is open to doubt.

The paired fins of osteolepiform rhipidistians, in particular those of Eusthenopteron and Sterropterygion have been cited as precursors of tetrapod limbs. Yet of the seven general conditions of the tetrapod limb outlined below, osteolepiforms satisfy only the first and part of the second.

1. An unpaired basal element (humerus or femur) supporting entire appendage.
2. Paired, subequal subbasal elements (ulna-radius and fibula-tibia) that are functionally joined distally.
3. Two primary joints (shoulder or hip and elbow or knee) in each appendage.
4. Distal bony elements (carpels or tarsals and digits) arising on postaxial side of appendage.
5. Pectoral and pelvic appendages alike in structure (i.e., structurally similar and with at least four or five primary segments).
6. Extensive muscular lobe that can be extended well below the ventral body wall.
7. Reduction or loss of dermal rays.

Dipnoans, on the other hand, meet all of the seven criteria. Thus, in the pectoral fin of a very young Neoceratodus, the radial associated with the second metapterygial segment is particularly large and articulates with the postaxial edge of the basal element (humerus) proximally and with the main axis distally. The relationships of this radial are exactly those of the ulna of a primitive urodele. Later, during growth the ulna of adult

Neoceratodus also possesses both pre- and postaxial radials, those along the dorsal edge being the larger. These large radials occur along the ventral fin edge in very young *Neoceratodus*, but during growth the pectoral fin rotates so that an original preaxial series becomes postaxial (the ulna, carpels and digits of urodeles are assumed to represent preaxial radials that are in the rotated postaxial position). In the pelvic fin the preaxial radial series remain in their original orientation (as in the tetrapod hind limb) thus the fin is a mirror-image of the pectoral. The coelacanth pectoral fin also appears to have rotated during ontogeny so that the original preaxial edge is postaxial in position.

The pectoral and pelvic appendages of lungfishes and tetrapods are of equal size and the muscular lobe of the lungfish, like the muscles of the tetrapod limb, reaches to the extremity. Moreover, the muscles are segmented in contrast to the non-segmented muscles in actinistians, actinoperygians and chondrichthyans.

Turning now to skull-roofing bones; most workers have concluded that the tetrapod pattern can most easily be derived from that of an osteolepiform rhipidistian. In order to accomplish this transformation, they have imagined that either a transverse series of scale bones (extrascapulars) have moved forward onto the neurocranium or that there was a backward movement of the parietals relative to the underlying neurocranium. In either case the arguments are not about observed pattern, but rather about untestable process. To us the osteichthyan dermal roofing bones appear to form two distinct patterns. The more general is that found in actinopterygians, osteolepiforms, porolepiforms and actinistians in which the paired frontals and parietals form the major constituents of the skull roof, the parietals reach the posterior limits of the neurocranium and the pineal foramen is situated between the frontals. The alternative pattern is seen in dipnoans and primitive tetrapods in which the eye lies opposite the junction between two pairs of large bones, there is a cluster of five bones (behind the two pairs of large bones) covering the temporal region and the pineal foramen lies between or just anterior to the parietals.

The palate of *Eusthenopteron* is supported by a long hyomandibula and is movably articulated with the braincase. The palatal surface is covered with granular teeth and a few large fangs. It is generally agreed that this type of palate is more likely to have been the tetrapod ancestral condition than that found in lungfishes which is fused to the neurocranium and supports large grinding toothplates. However, a similar palatal construction to that in *Eusthenopteron* (apart from the absence of fangs) is met with in actinopterygians while the primitive lungfish *Griphognathus* shows a palate in which the dermal bones form a pattern comparable to that in *Eusthenopteron* and these bones are covered with granular teeth. The type of palate seen in *Eusthenopteron* and actinopterygians is considered primitive for osteichthyans. On the other hand, in lungfishes and tetrapods, the pterygoids are large, joined in the midline, and exclude the parasphenoid anteriorly from the roof of the mouth. Furthermore in both the palate is fused to the neurocranium and the small hyomandibula (stapes of tetrapods) plays no part in jaw suspension.

In Recent lungfishes, the notochord is unconstricted and there is a single pair of cartilaginous arcualia above and below it. In *Neoceratodus* there are additionally, interventrals in the anterior thoracic region and interdorals in the tail. In the Devonian rhynchodipterids and in *Chirodipterus*, however, there are spool-shaped centra, which as in amniotes, are presumed to be made up of cartilage bone. The vertebral column of *Eusthenopteron* is similar to that of *Ichthyostega* and temnospondyls in which the pleuro- and intercentra are believed to represent endochondral ossifications of the cartilaginous arcualia similar to those in many fossil actinopterygians (*Caturus*, *Osteorhachis*). Thus, *Eusthenopteron* exhibits the generalized osteichthyan condition (in which centra are absent, Gardiner, 1983).

One final reason for excluding lungfishes from tetrapod ancestry was that they were presumed to possess, single-headed pleural ribs, whereas tetrapods and *Eusthenoptheron* were said to have double-headed dorsal ribs.

In *Polypterus* there are two series of ribs: dorsal ribs in the horizontal septum and pleural or ventral ribs in the wall of the coelom, while teleosts may have a third series. The dorsal ribs of *Polypterus* are peculiar and related to lateral cartilages (themselves unique to polypterids), whereas the dorsal ribs of teleosts characterize the higher teleosts. In all other gnathostomes there is a single rib series (presumed to be pleural or ventral ribs). Furthermore, in many gnathostomes, the successive members of a single series of ribs may change their positional relationships along the length of the vertebral column. A particularly striking resemblance in the rib series of lungfishes and tetrapods is the gradient in the size of the ribs along the length of the column. In both the largest ribs are found immediately behind the head and they become progressively shorter posteriorly. Finally, the presumed incipient development of a double head on the rib of

Eusthenoptheron; a similar development is also found in various selachians, and although diapophyses are found in tetrapods and on the anterior vertebrae of a number of actinopterygians and lungfishes, their absence in Eusthenopteron must be judged primitive.

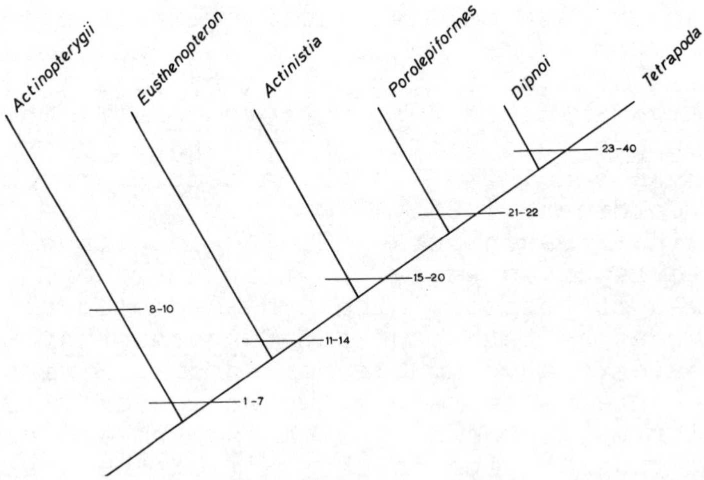

FIGURE 1. Character-state tree of Osteichthyes. Only a sample of characters which might be used to support this classification is included here. They are: (1) endochondral bone, (2) marginal teeth associated with dental arcades, (3) Lepidotrichia in the fins, (4) lung/swimbladder, (5) suprapharyngobranchials, (6) interhyal, (7) gular plates, (8) acrodin caps on teeth, (9) modified fin elements forming pelvic girdle, (10) ganoin, (11) teeth with enamel, (12) sclerotic ring of more than 12 segments, (13) exclusively metapterygial pectoral and pelvic fins, (14) submandibulars, (15) pectoral and palvic fins of equal lengths with long fleshy lobes, (16) preaxial side of pectoral fin endoskeleton rotated to postaxial position, (17) single, broad basibranchial, (18) last gill arch articulates with base of fourth arch, (19) inferior vena cava, (20) pulmonary vein, (21) cosmine pore canals enamel lined, (22) immobilization of intracranial joint, (23) choana, (24) two ball-and-socket joints in each paired

appendage, (25) reduction in ratio of dermal fin-rays to supports in paired appendages, (26) fleshy lobes of paired fins extending to tip, (27) muscles in paired fins segmented, (28) left and right pelvic girdles fused together to form pubic and ischial processes, (29) tetrapodous locomotion, (30) two pairs of dermal bones attached to the temporal region of the brain-case posterior to parietals, (31) larva in which the epidermis is ciliated, (32) glottis and epiglottis, (33) tetrapod-like neurohypophysial hormone, (34) structure of the lens proteins, (35) similar internal structure of the lung, (36) valves in conus separating gill-blood from lung-blood, (37) pterygoids joined in mid-line, (38) pulmonary circulation, (39) two-chambered auricle, (40) hyomandibular not involved in jaw suspension.

CONCLUSIONS

In the preceding analysis, we have tried to emphasize the shared derived characters of lungfishes and tetrapods, and have proposed alternative explanations of the similarities between osteolepiforms and tetrapods - that they are convergent or primitive.

To complete the analysis, a classification of bony fishes and tetrapods is presented in the form of a character state diagram (Fig. 1). The numbered characters are given in the accompanying legend.

From this analysis, *Eusthenopteron* turns out to be a very primitive fish and the lungfishes the closest relatives of tetrapods.

REFERENCES

Allis, E.P. (1932). The pre-oral gut, the buccal cavity and the buccopharyngeal opening in *Ceratodus*. J. Anat. 66, 659-668.

Dollo, L. (1896). Sur la phylogénie des dipneustes. Bull. Soc. Belge Géol. Paléont. Hydrol. 9, 70-128.

Gardiner, B.G. (1983). Gnathostome vertebrae and the classification of the Amphibia. J. Linn. Soc. (Zool.) 79, 1-59.
Jarvik, E. (1942). On the structure of the snout of crossopterygians and lower gnathostomes in general. Zool. Bidr. Upps. 21, 235-675.
Jarvik, E. (1972). Middle and Upper Devonian Porolepiformes from East Greenland with special reference to *Glyptolepis groenlandica* n.sp. Meddr Grønland 187, 1-295.
Watson, D.M.S. (1912). The larger Coal Measure Amphibia. Mem. Proc. Manchr lit. phil. Soc. 57, 1-14.
Watson, D.M.S. (1926). The evolution and origin of the Amphibia. Phil. Trans. R. Soc. Lond. (B) 214, 189-257.

CONSTRUCTIONAL MORPHOLOGY AND PHYLOGENETIC RECONSTRUCTION

WOLFGANG F. GUTMANN
Forschungsinstitut Senckenberg, Senckenberganlage 25, D-6000 Frankfurt a.M.1

ABSTRACT

Living organisms are hydraulic systems. Their form is determined by mechanical building elements which constitute a coherent frame of muscles, connective tissue, and fluid filled spaces.

All skeletal elements are developed in such constructions and borrow their form from the soft body. Consequently the emergence and phylogenetic transformation of skeletal elements have to be explained from construction principles of the hydraulic system. This is illustrated by two case studies.

THE CONSTRUCTIONAL PRINCIPLES OF THE SOFT BODY

Fossilized structures are valuable documents of the process of phylogeny and many of its radiating stages. As morphology and function of skeletal elements depend on the soft body and its architecture insight into the building principles of the living constructions will automatically improve our understanding of fossil structures and the phylogenetic changes the fossil record reflects in an all too fragmentary way. As all mechanism of life are firmly tied to liquid solutions enclosed in flexible membranes the animal machine has to be conceived as a hydraulic apparatus. This characterization is true as well for the single cells of protozans as for the metazoan constructions which usually possesses a body wall built of muscles, connective tissue, other soft structures, and a fluid or cellular filling (Gutmann 1972; Bonik et al. 1977).

To ensure functionality the body construction and the arrangement of its functional elements have to be organized according to strict lawlike construction principles which have been elaborated during the last two decades under the heading of hydrostatic skeleton (Clark 1964; Trueman 1975; Gutmann 1972, 1981; Gutmann and Bonik 1981; Vogel and Gutmann 1981). They can only be presented here in a very sketchy form.

As hydraulic systems show a strong tendency to assume a globular shape, every definite body shape and the maintenance of a complex architecture of the soft tissues depend on the presence of mechanical elements which help to suppress irregular deformations, ensure constancy of body shape, and control the pattern of orderly deformations during locomotion and other modes of behaviour. To keep the body architecture under control all parts of the body must be restrained by form-enforcing elements which have to be organized in all main directions of space. Furthermore, they are required to be integrated into a mechanically coherent whole which is closed

Proceedings of the 27th International
Geological Congress, Volume 2, pp. 45–52
PALAEONTOLOGY

so that the fluid filling of cavities and tissues cannot escape. While mechanical coherence is provided by a three-dimensional grid of connective tissue fibres, epithelia and their underlying supporting structures contain the filling in the body. Muscles which are constantly forming components of such constructions are required to be arranged in such a way that antagonistic interrelationships between separate or interwoven portions of the contractile system are guaranteed. While the internal relations of the body architecture are governed by these biomechanical principles of the soft tissues the general outer body shape is determined by the demands of the propulsion mechanisms.

If such machine-like constructions become subject to random mutations phylogenetic transformation can only follow certain pathways on which the internal and external requirements of the living apparatus and of the environment are fulfilled. In contrast to the widely held conceptions of selection as an external force selective constraints, which work on the organisms, have to be conceived as determined to a high degreee by the constructional and functional interrelationsships and the mutual dependences of organs and structures of the living organism itself. Consequently adaptation to external requirements can only be properly explained from the prior knowledge of the organismic construction and its possible transformations over functionally feasible intermediate stages (Gutmann and Peters 1973).

Because organisms are machine-like energy converting systems competition between them will result in a strong optimizing and economizing pressure as part of selection. Therefore, all organs which lost their function will shrink and become reduced, because the matter and energy for useless structures is selectively detrimental. In the context of this organism-centred evolutionary theory, which differs in essential aspects from the synthetic theory, a revised conception of selection with internal, external, and energetic constraints is proposed for utilization in reasonable attempts to reconstruct phylogenetic transformation series. From this follows that phylogenetic transformations have to be reconstructed in the form of theoretical models for morphological transformations canalized by internal organismic and mainly biomechanical constraints and external environmental limiting factors. Morphoclines of fossil and recent structures which are set up to represent the course of phylogeny can only be considered legitimate phylogenetic models if they can explicitely explain the selective advantages and the economizing effect which occured in the course of the transformations. Furthermore all intermediate stages of the gradualistic transformation have to be reconstructed as functionally feasible organisms.

As the body architecture of metazoans is mainly governed by the internal biomechanical and external behavioural and locomotory principles the models describing and explaining the emergence and the transformation of the morphology of major systematic groups has to be based on these. While the basic hydraulic machinery of animals does not require skeletal structures for its function, the form and function of skeletal formations attached to or integrated into the soft body can only be adequately analysed from the organization of the soft body. As there can be no doubt that the skeletal structures borrow their shape from the soft body and the arrangement of form-controlling elements the emergence and the change of skeletal parts must be viewed from the understanding of the soft body. This can be demonstrated for the chordates and the molluscs.

CASE STUDY 1: THE DEVELOPMENT OF THE VERTEBRATE BAUPLAN

Vertebrates are typical coelomates that possess a coelomic cavity which contains the internal organs. In contrast to the fully developed vertebrates, which are equipped with skeletal structures, their forerunners must have been worm-like animals. This conclusion is very reliable as in a sklereton-free coelomate the coelomic pouches are

FIGURE 1

Transformation sequence leading from a coelomic worm-construction to the craniote bauplan. A. Elongate hydraulic construction with fluid-filled cavities (D = gut, M = external muscle system). – B. Emergence of the body axis (Ch = notochord) which ensures constancy of length and allows reduction of all non-longitudinal muscles. Some fluid-filled spaces in the somites are still present (Lh). – C. Early craniote stage, all fluid-filled cavities in the somites have disappeared. – D. Emergence of skeletal structures (Sch = scales, Kn = skeletal rods at the body axis). They are integrated into a preexistent soft body construction. They serve to suppress disadvantageous deformations.

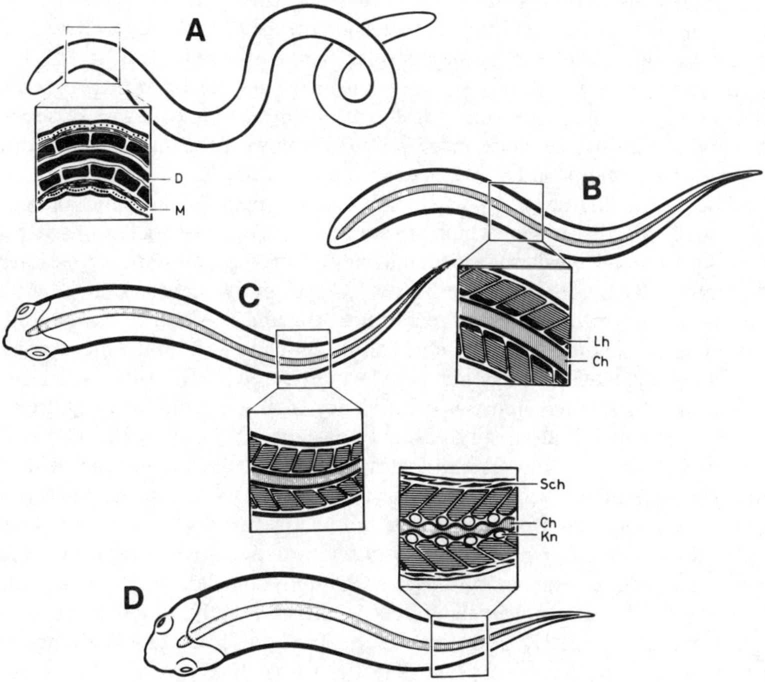

fluid filled cavities which serve as a part of the motoric apparatus. The elongate body shape of the worms is enforced by the requirements of locomotion. In primitive free living metazoans locomotion can only be effected by waves, either peristaltic or sinusoidal, travelling over the body.

When the worm-like forerunners of vertebrates turned to sinusoidal swimming movements they had to hold the body constant in length as only the sideways excursions could be utilized for propulsion. Suppression of length changes is ensured by muscles that brace the body while the longitudinal muscles cause the wriggling movements. In this situation the emergence of other structures which serve the same function and are energetically less expensive than muscles must provide an enormous selective advantage. In the nematodes (and other ashelminths) this problem is solved by a cuticle that prevents swelling of the worm which thereby is indirectly prevented from shortening. In the evolutionary path leading to the chordates the formation of a flexible rodlike structure, the notochord, could render the apparatus more effective. The primitive body axis is itself a hydraulic element as it consists of a fibre sheeth that keeps a filling of cells and fluid in its place. As this organ resists shortening it can hold the whole body constant in length. However, this mechanical effect is dependent on the integration of the body axis in the mechanical frame of the soft body. Only metameric worms could offer sufficient internal tissue and muscle bracings to hold the body axis in the middle of the tube-like body. While the axis was formed in the longitudinal mesenterial wall above the gut the transverse dissepiments with their muscles provided the transverse tensile structures to hold the axis in its place.

The economizing effect resulting from the emergence of the body axis was achieved in an indirect way. In the step by step development of the axis transverse and circular muscles could be reduced as they were no longer needed to brace the body and suppress shortening. At the same time the longitudinal muscles could increase in thickness and could therefore generate more propulsive power. During this transformation parts of the dissepiments had to persist as myosepta that linked the somitic muscles with the body axis. Thereby the persistence of metamerism in chordates and craniotes was enforced. From this model which is functionally feasible in all stages it becomes clear that the first and persisting skeletal structure of chordates and vertebrates is entirely determined by the interrelationships of the soft-body architecture.

When in the course of vertebrate evolution stiff and fossilizable skeletal elements arose they could only modify a mechanically complete soft body machine and improve its efficiency. Skeletal elements must be considered as structures which in consecutive steps cause a further increase of efficiency of the motoric activity. In addition to the body axis which had already ensured constancy of body length and continues to do so in all stages of craniotes imbricating rhombic scales and internal skeletal rods (ribs and processes of the vertebrae) became arranged in a way to suppress those deformations and movements of the muscle connective tissue system which could not be profitable for locomotion. Consequently the functional interpretation of skeletal structures and their change in phylogeny can only be derived from an adequate understanding of the biomechanical frame of the soft body which in all basic bauplan constructions is entirely autonomous in the absence of skeletal elements. Only during later stages of vertebrate phylogeny the whole apparatus can become subject to modifications for which skeletal structures are essential.

CASE STUDY 2: THE EMERGENCE OF THE MOLLUSC BAUPLAN

While in the evolutionary pathway leading to the craniotes the skeletal structures developed as stiff rods or scales within the connective tissue, skeleton-formation in molluscs is achieved by the emergence of outer skeletal plates to which the soft body remained attached. The key changes of body architecture and function connecting typical coelomates with the primitive mollusc construction can be summed up as follows.

As worm-like coelomates are very versatile organisms they are also capable of creeping on the substrate. This mode of locomotion is achieved by peristaltic movements. Because the ventral side of the body has to be held flat to gain contact with the ground, densely arranged muscle bearing dissepiments, subdividing the body cavity, were required in the incipicient stages. Unless the internal muscle bracing could control the cross-section of the body and adjust it to the contour of the substrate, effective creeping would have been prevented from developing. Therefore, a metameric ancestor must be postulated for molluscs.

FIGURE 2

Formation of the mollusc bauplan. A. Worm-like metameric ancestor which could change its cross-section for creeping on the ground. – B. Transitional stage with thickened foot. – C. Formation of a series of skeletal plates. – D. Conchiferen stage with fused shells carried on the dorsal side.

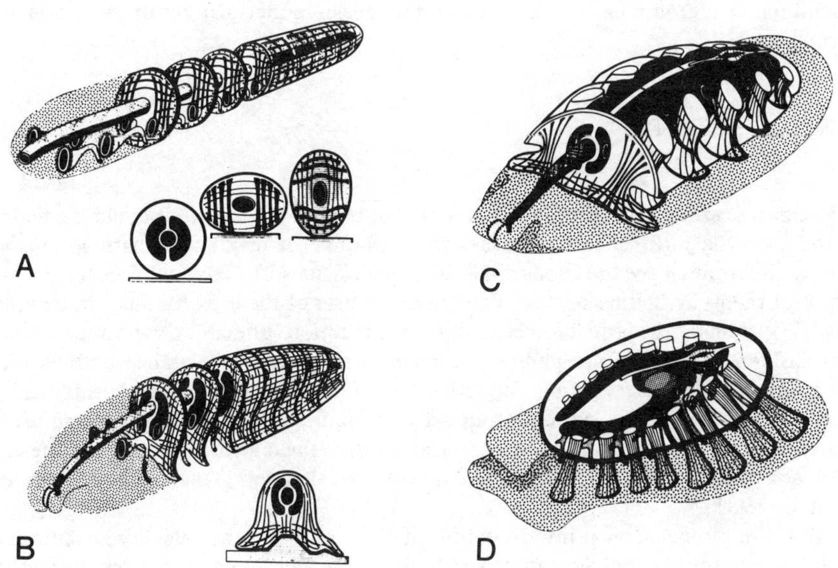

Improvement of creeping and adhesion of the body to the substrate could render scraping movements of the developing radula more effective. As a result of these selective advantages a thick muscular foot developed on the ventral side.

At the same time the coelom and the other internal organs were shifted to the dorsal body regions. While the peristaltic deformations of the body became more and more restricted to the flat ventral foot, the coelomic cavity gradually lost its function as a fluid filling and the dorsal parts of the body were held still. While in the first step suppression of the deformations of this region was effected by the bracing influence of the muscle system, the formation of stiff skeletal structures could reduce the amount of energy and material needed to control the shape of the dorsal side of the body. This led to the formation of skeletal plates which, beyond their economizing effect, could also serve as protective devices in the slowly browsing animals.

The first stage of shell formation must have consisted in the emergence of a series of shell plates. This is documented by the chitons which in certain respects represent a more primitive stage of molluscan phylogeny than the conchifera. In the primitive molluscs creeping over irregular ground required bending of the body which was made possible by separate but mobile shell plates. So some properties of worms coexist with mollusc features in chitons.

In contrast to the chiton stage the conchiferan molluscs which represent an advanced level acquired a fused shell which contains the viscera in a position shifted further to the dorsal side. Integration of separate plates into an unique shell was rendered possible by the formation of a much thicker foot which could absorb even major irregularities during movement over uneven ground.

As in the vertebrates it should not be forgotten that even in later stages of molluscan phylogeny after the skeleton is firmly established, the shape of the animal and the form of its shell are ensured by the bracing of muscles. Derivation of the diverging conchiferan subgroup has to start from the basic model just outlined (Vogel and Gutmann 1980).

GENERAL RESULTS

In the case studies just given (and in others published elsewhere) it could be demonstrated that the most ancestral stages of the bauplan of major systematic groups are entirely determined by the biomechanical frame of the soft body and thus by the properties of the hydrostatic skeleton. Even in the course of the establishment of complex skeletal systems the basic biomechanical construction principles continue to exert their influences on the morphological organization. Since skeletal structures must never impair active locomotion and other essential behavioural mechanisms, and as larger skeletal elements cannot form in body portions which are not held immobile by muscle action, the biomechanical organization and the motoric activity of the soft body are essential in determining where skeletal elements can emerge (Vogel and Gutmann 1981).

From the biomechanical interpretation of the soft body the selective advantage of newly developing skeletal structures and their further development can be inferred. As skeletal elements form stiff structures in the context of the soft body they will help to suppress disadvantageous and energy consuming deformations of the soft body. At the same time motoric activity, indispensable for locomotion and other modes of behaviour, remains unimpaired and will automatically gain in efficiency. In many cases protective functions of the skeleton will arise pari passu with the economizing effect on the motoric functions.

As the audience might not be familiar with the reconstructive procedure it should be emphasized that the hydraulic principles and the interrelationship of its functional components, which did not attract sufficient attention of morphologists, are based on very simple and valid physical laws and on reliable physiological properties of the building materials.

The conclusions drawn from the constructional principles and simple mechanical laws are fundamental for the reconstruction of possible phylogenetic pathways. Application of these principles allows the formulation of theories that explain the sequence of modifications that render the organization and the development of fossil and recent animals plausible. The pertaining methodology for phylogenetic reconstructions has been developed during the 1970ies by several authors (Peters and Gutmann 1971; Bonik et al. 1977; Vogel 1979; Bock 1979, 1980). Models for other fossil organisms have been published elsewhere. In order to eliminate from phylogenetic considerations those derivations and morphoclines which are based on similarity alone they have to be replaced by theories in which the constructional principles of the soft body and the requirements of locomotion and some aspects of behaviour have to be explicitely stated. The theoretical models tracing transformation series over functionally feasible intermediate stages must of course lead to fossil and recent animals.

REFERENCES

Bock, W. J. (1979). The Synthetic Explanation of Macroevolutionary Change. A Reductionistic Approach. Bull. Carnegie Mus. Nat. Hist. 13, 30 - 69.

Bock, W. J. (1980). The Definition and Regcognition of Biological Adaptation. Amer. Zool. 20, 217 - 227.

Bonik, K., Grasshoff, M. and Gutmann, W. F. (1977). Funktion bestimmt Evolution. Umschau 77 (20), 657 - 668.

Chapman, G. (1975). Versatility of Hydraulic Systems. J. Exp. Zool. 194, 249 - 270.

Clark, R. B. (1964). Dynamics in Metazoan Evolution. Clarendon Press, Oxford, pp. 1 - 313.

Gutmann, W. F. (1972). Die Hydroskelett-Theorie. Aufsätze u. Red. senckenb. naturf. Ges. 21, 1 - 91.

Gutmann, W. F. (1974). Die Evolution der Mollusken-Konstruktion: Ein phylogenetisches Modell. Aufsätze u. Red. senckenb. naturf. Ges. 25, 1 - 84.

Gutmann, W. F. (1981). Relationships between invertebrate phyla based on functional-mechanical analysis of the hydrostatic skeleton. Amer. Zool. 21, 63 - 81.

Gutmann, W. F. and Bonik, K. (1981). Kritische Evolutionstheorie. Gerstenberg, Hildesheim, pp. 1 - 227.

Gutmann, W. F. and Peters, D. S. (1973). Konstruktion und Selektion: Argumente gegen einen morphologisch verkürzten Selektionismus. Acta Biotheoretica 22 (4), 151 - 180.

Peters, D. S. and Gutmann, W. F. (1971). Über die Lesrichtung von Merkmals- und Konstruktions-Reihen. Z. zool. Syst. Evolutionsforsch. 9 (4), 237 - 263.

Trueman, E. R. (1975). The Locomotion of Soft-Bodied Animals. Arnold, London, pp. 1 - 200.

Vogel, K. (1979). Efficiency of biological constructions and its relation to selection and rate of evolution (General remarks). Palaeogeogr., Palaeoclimatol., Palaeoecol. 28, 315 - 319.

Vogel, K. and Gutmann, W. F. (1980). The derivation of Pelecypods: role of biomechanics, physiology and environment. Lethaia 13, 269 - 275.

Vogel, K. and Gutmann, W. F. (1981). Zur Entstehung von Metazoen-Skeletten an der Wende vom Präkambrium zum Kambrium. Festschr. wiss. Ges. J.-W.-Goethe-Univ. Frankfurt/M. 517 - 537. Franz Steiner, Wiesbaden.

COMPARATIVE ANALYSIS OF THE BENTHOS ECOLOGY IN ANCIENT BASINS

HECKER, R. TH., OSIPOVA, A.I., BELSKAYA, T.N.
(Paleontological Institute of The Academy of Sciences of the USSR, Mowcow, USSR).

As long ago as in the 1930's R. TH. Hecker showed in his first works of methodic character, that joint work of lithologists and paleontologists is necessary for the ecological study of ancient basins. At the same time he proposed a research program, a special attention being paid to the study of fauna changes in some biotopes through geological time. The solving of the problems of this range became possible, however, only by means of combined lithological and paleoecological investigations of a number of phanerozoic basins. These works were accomplished by the students of the Laboratory of marine fauna paleoecology in the Paleontological Institute of the Academy of Sciences of the USSR, and in some other scientific-research institutes. The results of these researches are elucidated in a number of monographs and reports, and reviewed in the book by A.I. Osipova (1980) and in some articles (Osipova et al., 1983, a. oth.).

Zonal distribution of the main groups of benthic organisms inhabited the shelves of Silurian, Devonian and Early Carboniferous basins is considered in the present report. The records of Silurian basins in the East Baltic, Podolia and Southern Siberia, and of Devonian and Early Carboniferous ones in Siberia and on Russian platform are used as a base for comparison. These basins were different by their rectonic position, morphology and their connections with the ancient oceans. The processes of sedimentation, however, as well as environmental conditions of benthic life were similar enough on the shelves of Paleozoic seas, though such characteristics as shelf width, floor gradient, supply of terrigenous material and rate of

Proceedings of the 27th International
Geological Congress, Volume 2, pp. 53–67
PALAEONTOLOGY

sedimentation varied essentially. Therefore it is possible to compare ecology of the benthos, inhabited basins of different types, as it was shown by E.A. Ivanova (1962).

The Silurian East Baltic sea is considered to be a pericontinental basin adjacent to the peneplanated land. As a result calcareous sediments predominated in the shallow-water part of the basin. Three facies zones (belts) are distinct on the shelf, namely lagoons, inshore shoals and open shelf (Fig. I.A. Zones I-III). Off-shore deposits contained more argillaceous material and organic matter are interpreted as sediments of basin slope (with depth more than 90 m) and as that of more deep-water depression zone (zones IV and V) (Kaljo, 1978; Nestor, Einasto, 1982; Kaljo et al., 1983). Lagoon (I) with not normal marine salinity, where dolomitic sediments accumulated, was dwelt by a specific impoverished benthic assemlage: lingulids, ostracodes (with Leperditia and Herrmannina predominance) and eurypterids. Benthic organisms were abundant and diverse on the inshore shoals separated lagoon and open shelf (II). In the high agitated water stromatoporoids, tabulate and rugose corals settled in masses, as well as crinoids. All these animals found optimum living conditions here and formed bioherms sometimes. Abundance of animals of all these groups decreased on the open shelf. Brachiopods were more widely distributed - from lagoon to the depression zone, in the latter only lingulids dwelt. Brachiopods reached their top diversity in the open shelf zone (III). Ostracodes distribution was similar to that of brachiopods. Trilobites reached the highest diversity and abundance on soft substrates in the outer part of the open shelf and on the slope.

Distribution pattern of facies zones and of main benthos groups in the Silurian basin of Podolia was similar to those in East Baltic Silurian sea (Predtechensky et al., 1983). In some environments - with abnormal salinity and high agitated water or, on the contrary, on the soft sediments of the most deep part of the shelf, bivalve molluscs were abundant (M.R. Hecker, 1981). Such conditions were unfavourable for coelenterates, brachiopods and crinoids.

Early Silurian sea of Tuva depression (Southern Siberia; see Ivanova et al., 1964) was characterized by high terrigenous supply. Terrigenous material was accumulated mostly in the marginal zone. In the shallow-water zone (corresponding to open shelf of the East Baltic sea) argillaceous and biodetrital calcareous sediments were accumulated. The zone was inhabited by various brachiopods, trilobites, bryozoans etc. Brachiopods predominated here and were represented by Atrypa, Tuvaella, Spiriferids etc., as well as Schizophoria and Rhipidomella. Unlike the East Baltic and Podolian basins, coelenterates dwelt mostly on the off-shore shoals free of terrigenous material. They often formed bioherms, inhabited also by abundant crinoids.

Distribution pattern of facies zones and of benthos in the Late Devonian (Frasnian) seas was similar to those in the Silurian sea of Tuva. Most diverse brachiopod and ostracode assemblages inhabited the shallow-water zone (Fig. I, B, zone IIIa), and coelenterates settled on the off-shore shoals (IIIb zone). In brackish-water lagoons lingulids and ostracodes Leperditia dwelt. In the marginal zone brachiopods predominated, namely Athyris and Retzia, tolerant to the reduced salinity. Brachiopods Gypidula settled in masses on the off-shore shoals slopes. The relative deep-water zone (IV) was dwelt by small brachiopods Isopoma and Leiorhynchus, the latter is known to be an indicator of disturbed aeration.

The end of Frasnian is considered to be the time of one of the most important extinctions wasted the shallow-water marine environments (Boucot, 1975). Benthos was abundant, however, in Kuznetsk sea near the Frasnian-Famennian boundary, though diversity and abundance of stromatoporoids and rugose corals decreased abruptly, and brachiopods and bryozoans became less diverse (17 genera of brachiopods in Famennian sea instead of 25 genera in the Frasnian one). Coelenterates were substituted with calcareous algae on the off-shore shoals (Belskaya, 1960).

Another Fransnian basin studied in detail was that of the Main Devonian Field in the North-West of Russian platform. Facies zone relations in the basin (Fig. I,c) were different from those of the basins mentioned above. The Main Devonian Field sea was conterminous in the North-East with a vast coastal alluvial plain. This plain was the source of fresh water and of terrigenous material accumulated in the eastern marginal zone (I and II); salinity was reduced to a various degree. More off-shore, in the shallow-water zone (III) salinity was close to the normal marine one, and argillaceous and calcareous sediments were accumulated here. In the central part of shallow sea shoals were situated. Calcareous sediments accumulation often broke off here, and hard-grounds were formed. This habitat was characterized by a biocoenosis formed by attached animals (brachiopod Irboskites a.oth., bivalve mollusks, crinoids, edrioasteroids, auloporids), boring ones, brachiopods of anchor ecological type, etc. (Hecker, 1965, 1983). To the west of the shoal there was a vast gulf with increased salinity (III c), where calcareous-dolomitic and dolomitic sediments accumulated. Habitats with reduced salinity (marginal zone) and with increased one (the gulf) were inhabited by benthos similar impoverished: lingulids, ostracods, euryhaline articulate brachiopods with predominance of Ripidiorhynchus. When salinity became very high, only eurypterids and conchostraca survived. Abundant brachiopods (Ladogia, Schizophoria, Atrypa etc.) dwelt in the shallow water zone with salinity close to normal marine, but the number of genera was here half of that in the Frasnian sea of Kuznetsk basin. Stromatoporates and tabulate corals were not diverse, they did not form bioherms and had a limited distribution (except auloporids). Rugose corals were very rare too.

Bivalve mollusks, on the contrary, were widely distributed in the habitats with abnormal salinity.

Late Visean sea of the central part of the Russian platform was situated in the humid tropical climate area. Sandy, silty and argillaceous sediments rich in plant detritus and carbonaceous organic matter accumulated in marginal zone (Fig. I,D, zone I and II), adjacent to a bogged coastal plain with rich vegetation. In the more off-shore shallow-water zone (III) biodetrital calcareous sediments accumulated mainly. When shallowing shoals and low islands arose in the zone, with lepidophyte thickets and with lagoons where chemogenous micrograined calcareous sediments accumulated (IIID). Marginal zone (I and II) was inhabited by benthic organisms tolerant to salinity fluctuations, disturbed aeration and high turbidity. Diverse ostracodes, brachiopods (see below), rare and monotonous foraminifers, skeletonless mud-eaters, etc., dwelt here. In the more off-shore shallow-water zone brachiopods and foraminifers diversity was more high, but ostracod abundance reduced, bairdiids being dominant. The latest data on the Paleozoic basins of Western Europe and North America confirm the connection of Leperditia assemblages with the marginal zone environments, and of assemblafes with bairdiids predominance with the normal marine ones (Lethiers, Braun, 1983). Calcareous algae Calcifolium were distributed in masses. Various foraminifers and coelenterates (solitary and colonial rugose corals, syringoporids and chaetetids) predominated in the off-shore zone (300-400 km away from the shore-line) with its optimum normal marine conditions. But abundance and diversity of brachiopods decreased here. Only Striatifera were an exception, sometimes they formed banks. Ostracodes were very rare and of low diversity.

Comparison of Late Visean marine basin and more ancient ones shows evidently, that distribution patterns of benthic organisms have changed significantly in the time from Silurian to Early Carboniferous.

1. In Silurian and Devonian shelf basins biotopes with normal-marine salinity (quiet water with soft calcareous-argillaceous sediments as well as calcareous substrates dwelt by stromatoporoids, corals etc.) were inhabited by ostracode assemblages of high diversity. Impoverished assemblages, usually with predominance of leperditiid ostracodes, inhabited marginal environments only (with reduced or increased salinity). Visean marine basins were characterized by an inverted distribution pattern. Ostracodes were most various in the marginal part of Visean basin on soft calcareous-argillaceous and argillaceous-calcareous sediments, whereas their diversity and abundance decreased abruptly in the more off-shore shallow-water environments (Zanina, 1956). Such distribution pattern most likely has been resulted from some biotic interrelations. Early Carboniferous is known to be the time of small foraminifers flourishing and of their achievement of new habitats (Rauser-Chernoussova, Reitlinger, 1957). Foraminifers were widely distributed on the shelves, and ostracodes likely being their possible competitors for nutrition had been forced out towards the habitats unfavourable for foraminifers, i.e. the marginal parts of basins and, on the other hand, towards the most deep environments with reduced hydrodynamic activity.

2. Stromatoporates being so widely distributed in more ancient seas were absent in the Visean sea of the Russian platform. It was the result of their nearly completely extinction (as well as that of majority of tabulate corals) near the Devonian-Carboniferous boundary. Extraordinarly rare and obviously stunted stromatoporoids of the Clathrodictyonidae family (flourished during Silurian and Devonian) were found in the Serpukhovian of Donetz basin (Bogoyavlenskaya, 1982). Habitats of tabulate corals have been decreased: some representatives of this group (Thamnopora, Taxopora a.oth.) were sometimes abundant in the Silurian and Devonian seas in the habitats, where salinity was abnormal, though they were usually most abundant and diverse in the

normal marine environments. In the Visean sea tabulate coral settled only in the well aerated water of normal salinity on the pure calcareous sediments. Various rugose corals (Palaeosmilia, Lonsdaleia, Corwenia a.oth.) inhabited the same environments; the only exception were zaphrentoid rugose corals, capable to existing on soft argillaceous substrates in the quiet-water habitats.

3. Brachiopods were distributed all over the shelf of the Carboniferous sea. But their diversity became less towards off-shore zone, unlike to that in Silurian and Devonian seas. The number of taxa increased tolerant to the fluctuations of environmental factors characteristic of the marginal zone. Besides lingulids, inherited from the more ancient basins, as well as camarotoechiids and athyridids, some productids dwelt in these habitats forming three oligomixt (opportunistic) communities, i.e. Productus redesdalensis, Fluctuaria and Eomarginifera ones. They were associated by Schizophoria and Rhipidomella, the representatives of ancient genera inhabited the outer part of the open shelf in Silurian and Devonian seas. These brachiopods were forced out by the representatives of brachiopod genera, which appeared in the Carboniferous (Semiplanus, Gigantoproductus, Antiquatonia) and occupied the habitats more favourable for benthic animals (Osipova et al., (1971).

Many Carboniferous brachiopods were better adapted for dwelling in the various habitats than the more ancient representatives of the same ecological types - in particular due to the stronger valve articulation (the hinge of pentameroid brachiopods, for instance, was not strong enough - see Modzalevskaya, Nikiforova, 1980), and to the more "economic" construction of "pontoons" of free-lying brachiopods, inhabited soft substrates (Ivanova, 1962, 1971).

4. The number of trilobite genera were reduced abruptly in the Carboniferous. In the Tournaisian sea of the Kuznetsk basin trilobites still dwelt on the open shelf, but in Visean and Serpukhovian seas of the Russian platform their habitats were limited with the quiet-water environments of marginal zone with soft fine-grained calcareous and argillaceous sediments.

The comparative analysis gives an opportunity to characterize (as a preliminary) ecogenetic trends of main groups of benthic animals, inhabited the shelves of Silurian, Devonian and Carboniferous basins and show some characteristic features of ecogenetic process. During the evolution flourishing of a group its representatives inhabited (as it is known for trilobites, stromatoporoids, tabulate corals, ostracodes, some brachiopods etc.) a wide spectrum of habitats, up to the environments with abnormal salinity, but they predominated in the normal marine optimum environments. Through time (and especially after great extinctions) the other branches of benthic animals entered the flourishing phase, and the habitats of more ancient ones became limited. The most part of ancient genera was forced out towards the less favourable environments on the shelf. Some of them reached high abundance in the new habitats (e.g. Carboniferous ostracodes), and some became extinct (e.g. trilobites). More rarely the representatives of ancient groups continued dwelling in the optimum habitats, but their abundance and diversity were reduced (tabulate corals, for instance).

A large number of hypotheses was suggested in order to explain ecogenesis and extinctions phenomena. Some time ago H. Nestor (1977) reviewed them, when considering the stromatoporate ecogenesis. Basing on the data got by Rh. W. Fairbridge (1967) on the increasing of oxygen content in hydro- and atmosphere due to development of land plants during the Devonian, Nestor gave an assumption, that profound aeration of oceanic waters had took place at the same time and the vertical circulation had been arose similar to the recent one.

We have paid attention to another factor connected with the land vegetation evolution, namely on supply of humic organic matter in the Carboniferous marine basins (Osipova et al., 1971, 1983). This factor was not significant for the Early Paleozoic seas where sapropelite organic matter predominated (Volkova, Lapo, Makedonov, 1979), but its influence was especially appreciable in the marginal zone of the Carboniferous sea basin. New nutrition resources appeared here; increased abundance of ostracodes and advancement of some brachiopods towards near-shore habitats were likely the result of this. Disturbed oxygenation was on the other hand resulted from large quantity of organic matter. This together with supply of fresh water and terrigenous material influenced negatively on the coelenterates and caused their migration towards the off-shore habitats.

These conclusions must be verified by comparative analysis of benthos distribution in other ancient basins, and especially in the Middle and Late Carboniferous and Permian ones. Deposits of these ages are widely distributed on the territory of the USSR.

REFERENCES

Belskaya T.N. 1960. Late Devonian sea of Kuznetsk basin, its historical development, population and sediments. Moscow, Nauka, p. 184. (Trudy Paleontol. inst. Adad. nauk SSSR, vol. 82) (in Russian).

Bogojavlenskaya O.V. 1982. Stromatoporata of Late Devonian and Early Carboniferous. -Paleontol. zhurn., N 1, p. 33-38 (in Russian).

Boucot A.J. 1975. Evolution and extinction rate controls. Amsterdam-Oxford-N.Y., Elsevier Co,

Fairbridge Rh.W. 1967. Carbonate rocks and paleoclimatology in the biochemical history of the Earth. -In: Carbonate rocks. Origin, occurrence and classification./G.V. Chilingar et al., eds. Amsterdam etc., Elsevier C°, p.

Hecker M.R. 1981. Ecology and facies connections of bivalve molluscs in Silurian and Early Devonian basins of Podolia. - Paleontol. zhurn., N 1, p. 15-24 (in Russian).

Hecker R.F. 1965. Introduction to paleoecology. N.Y. etc., Elsevier C°, p. 165.

Hecker R.Th. 1983. Taphonomical and ecological peculiarities of fauna and flora of Main Devonian field. Moscow, Nauka, p. 141. (Trudy Paleontol. inst. Akad. nauk SSSR, vol. 190) (in Russian).

Hecker R.Th., Osipova A.I. 1970. Regularities in distribution and changes of fauna in Visean and Early Namurian epicontinental seas of the Russian platform. - In: Compte rendu VI Congr. intern. stratigr. et geol. Carbonifere, Sheffield, 1967. Vol. III. Maastricht, E. van Aelst, p. 913-922.

Ivanova E.A. 1962. Ecology and development of brachiopods in Silurian and Devonian of Kuznetsk, Minusinsk and Tuva basins. Moscow, Izd. Akad. nauk SSSR, p. 152. (Trudy Paleontol. inst. Akad. nauk SSSR, vol. 88) (in Russian).

Ivanova E.A. 1971. Introduction to the study of Spiriferids. Comparative morphology. Moscow, Nauka, p. 105. (Trudy Paleontol. inst. Akad. nauk SSSR, vol. 126) (in Russian).

Ivanova E.A., Belskaya T.N., Tchudinova I.I. 1964. Environmental conditions of marine Silurian and Devonian fauna in Kuznetsk, Minusinsk and Tuva basins. Moscow, Nauka, p. 226. (Trudy Paleontol. inst. Akad. nauk SSSR, vol. 102) (in Russian).

Kaljo D. 1978. On the bathymentric distribution of graptolites. - Acta paleontol. Polonica, vol. 23, p. 523-531.

Kaljo D.L., Einasto R.E., Klaamann E.R. et al., 1983. An ecological model of the East Baltic. Silurian basin. - In: Problemy ekologii fauny i flory drevnikh basseinov. Moscow, Nauka, p. 43-61. (Trudy Paleontol. inst. Akad. nauk SSSR, vol. 194) (in Russian).

Lethiers F., Braun W.K. 1983. Les Ostracodes marquere des paleoenvironmentes paleozoiques. - In: Premier Congres International de Paleoecologie. Resumes. Lyon, p. 094.
Modzalevskaya T.L., Nikiforova O.I. 1980. Ecological and taphonomical analysis of the brachiopods of the Podolian Wenlock. -Yezhegodnik Vsesoyuzn. Paleontol. Obshch., vol. 23, p. 159-169. Leningrad, Nauka (in Russian).
Nestor H. 1977. On the ecogenesis of the Paleozoic stromatoporoids. -Mem. Bureau Rech. geol. min, N 89, p. 249-254.
Nestor H., Einasto R. 1982. Application of shelf and slope concepts to the Silurian Baltic basin. - In: Ecostratigraphy of the East Baltic Silurian Tallinn, Valgus, p. 17-25.
Osipova A.I. 1980. On the history of paleoecology in Russia and in the USSR. Moscow, Nauka, p. 65. (Trudy Paleontol. inst. Akad. nauk SSSR, vol. 185) (in Russian).
Osipova A.I., Belskaya T.N. 1970. Use of combined palaeoecological and lithological investigation for analysis of facies and of the process of sedimentation. - In: Compte rendu VI Congr. intern. stratigr. et geol. Carbonifere, Sheffield, 1967. Vol. III. Maastricht, E. van Aelst, p. 1293-1304.
Osipova A.I., Belskaya T.N. 1977. On the investigation of the Carboniferous of the Russian platform as a sedimentation model for epicontinental sea of humic zone. - Lithologiya i polezn. iskop., N 2, p. 48-62 (in Russian).
Osipova A.I., Belskaya T.N., Fomina E.V. 1975. Paleoecology of benthos in Early Carboniferous seas of the Moscow syneclise. - In: Paleogeografia SSSR, vol. 2 Moscow, Nedra, p. 104-108 (in Russian).
Osipova A.I., Belskaya T.N., Hecker R.Th. 1983. Investigations on ecology and on evolution of main benthic groups in the Carboniferous seas of the Russian platform. - In: Problemy ekologii fauny i flory drevnikh basseinov. Moscow, Nauka, p. 98-119. (Trudy Paleontol. inst. Akad. nauk SSSR, vol. 194)

(in Russian).

Osipova A.I., Hecker R.Th., Belskaya T.N. 1971. Some features of the fauna distribution and changes in the Late Visean-Early Namurian epicontinental seas of the Moscow syneclise. - In: Sovremennye problemy paleontologii. Moscow, Nauka, p. 279-293. (Trudy Paleontol. inst. Akad. nauk SSSR, vol. 130) (in Russian).

Predtechensky N.N., Koren T.N., Modzalevskaya T.L., Nikiforova O.I. et al. 1983. Sedimentation cycles and ecological succession of faunal assemblages in the Silurian of Podolia. - In: Problemy ekologii fauny i flory drevnich basseinov. Moscow, Nauka, p. 61-74. (Trudy Paleontol. inst. Akad. nauk SSSR, vol. 194) (in Russian).

Rauser-Chernoussova D.M., Reitlinger E.A. 1957. On the Foraminifera evolution during Paleozoic and on their stratigraphical value. - Izvestia AN SSSR, ser. geol., N 11, p. 103-124. (in Russian).

Sorokin V.S. 1978. The stages of the development of the north-western part of the Russian platform during Frasnian. Riga, Zinatne, p. 282. (in Russian).

Volkova I.B., Lapo A.V., Makedonov A.V. 1979. The changes in quantity and in composition of organic matter during geologic time. - In: Paleontologia i evoluzia biosphery. Abstracts of XXV sess. Vsesoyuzn. Paleontol. obshch. Leningrad, p. 9-10 (in Russian).

Zanina I.E. 1956. Ostracodes of Visean stage of Moscow basin. - In: Mikrofauna SSSR. Sborn. VIII. Leningrad, Gostoptekhizdat, p. 185-293. (Trudy VNIGRI, n.ser., vyp. 98) (in Russian).

FIGURE 1. (Appears overleaf.) Zonal distribution of the main groups of benthic animals in the Silurian, Devonian, and Carboniferous marine basins. A - Silurian sea of the East Baltic; shelf breadth about 150 km (after Kaljo, 1978; Nestor, Einasto, 1982; Kaljo et al., 1983); B - Late Devonian sea of the Kuznetsk basin; shelf breadth about 200 km (after Belskaya, 1960); C - Late Devonian sea of the Main Devonian Field of the Russian platform; shelf breadth more than 600 km (after Hecker, 1983 a. oth., and and Sorokin, 1978, a. oth.); D - Visean sea of the Moscow syneclise; shelf breadth more than 700 km (after Osipova, Belskaya, 1970, 1977).
Sediments: 1 - red-beds; 2 - sands and silts; 3 - sandy and calcareous sediments; 4 - clays; 5 - alternation of calcareous and argilaceous sediments; 6 - calcareous-argilaceous; 7-9 - calcareous sediments; 7 - biodterital, shelly, etc.; 8 - biohermic; 9 - micritic; 10 - dolomites; 11 - coal.
Benthos: 1 - stromatoporoids; 2 - tabulate corals (except auloporoids; 4 - rugose corals (a, colonial, b, solitary); 5-12 - brachiopod assemblages; 5 - lingulids; 6 - with predominance of pentamerids or Gypidula; 7 - with predominance of Retzia, Athyris; 8 - Isopoma, Leiorhynchus a. oth.; 9 - with predominance of Ripidiorhynchus; 10 - "Camarotoechia", small productids, chonetids; 11 - striatifera; 12 - assemblages of high diversity; 13-16 - ostracode assemblages; 13 - impoverished with Herrmannina and Leperiditia; 14 - normal-marine assemblages of high diversity; 15 - assemblage of the marginal zone of Early Carboniferous basins; 16 - with predominance of bairdiids; 17 - trilobites; 18 - eurypterids; 19 - foraminifers; 20 - mangrove-like plants. The signs are larger in the zones where abundance and diversity of a group were more high.

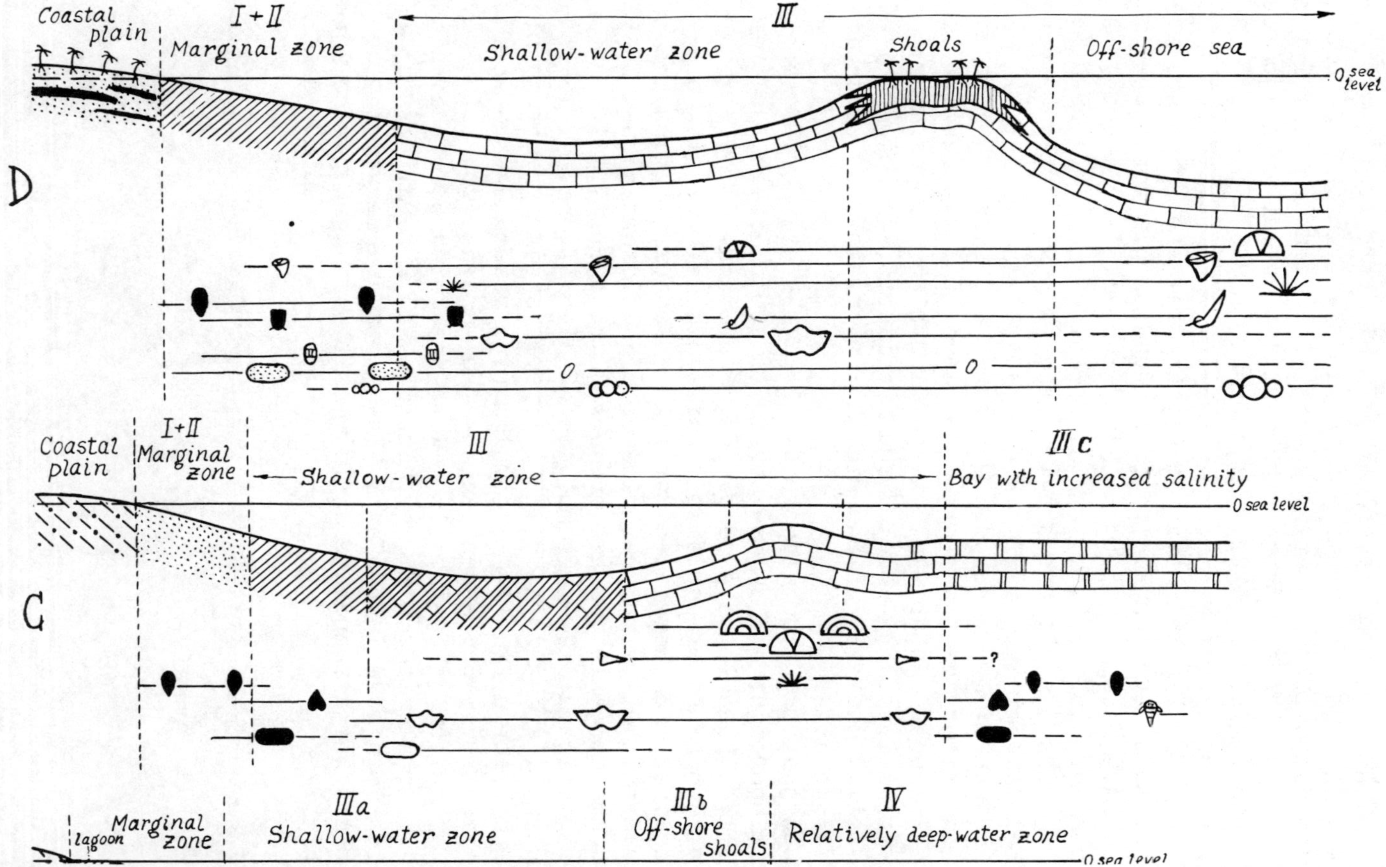
Coastal plain
I+II
Marginal zone
III
Shallow-water zone
Shoals
Off-shore sea
0 sea level
D
Coastal plain
I+II
Marginal zone
III
Shallow-water zone
III c
Bay with increased salinity
0 sea level
C
lagoon
Marginal zone
III a
Shallow-water zone
III b
Off-shore shoals
IV
Relatively deep-water zone

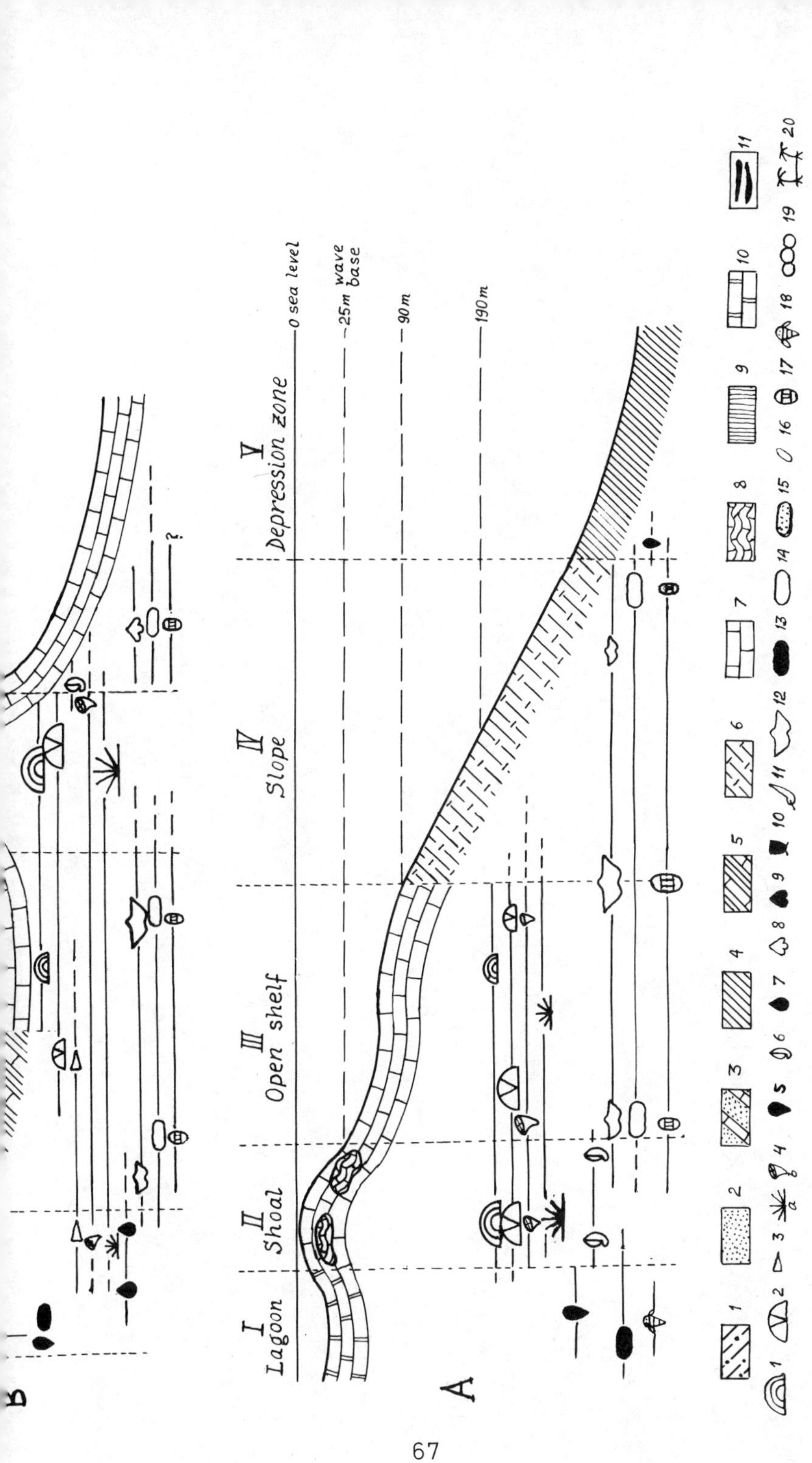
Б
A
I Lagoon
II Shoal
III Open shelf
IV Slope
V Depression zone
0 sea level
25m wave base
90m
190m
?
1 2 3 4 5 6 7 8 9 10 11
1 2 3 4 5 6 7 8 9 10 11 12 13 14 15 16 17 18 19 20

ECOSYSTEMS OF THE EAST BALTIC SILURIAN

KALJO, D., KLAAMANN, E., NESTOR, H.
Geological Institute of the Estonian Acad. of Sci., Tallin, USSR.

Palaeontology has borne relations with geology long since, mainly through stratigraphy, however, sometimes at a rather elementary level (lists of recorded species). Palaeoecologic approach, the application of the concept of ecosystem to the study of geological past in particular, have considerably widened mutual relations between these sciences, and in several fields of geology palaeontological argumentation has become decisive. E.g., complex study of sedimentary basins, acknowledged as an essential trend in modern geology (Sokolov, 1978) is impossible without palaeontological study of palaeoecosystems.

Nature and composition of a fossil ecosystem is insufficiently ascertained so far. Proceeding from the concept V. Krasilov (Krasilov, 1970) has laid into the term "palaeoecosystem" we acknowledge two possible interpretations of a palaeoecosystem of different geological, particularly stratigraphical sense. The first one which is based on the internal unity and mutual dependence of the development of environment and biota provides a basis for local lithostratigraphy, and the second one which considers ecosystem as a biosphere, and its development as a succession of palaeoecosystems (palaeobiospheres) serves as a basis for chronostratigraphy (Kaljo, 1982). If the former orients us to the discrete character of subdivisions, or, in other words, to the search of their "natural" boundaries, then the latter points out a necessity for the correlation of geological events, evolution of organic world, etc.

Proceedings of the 27th International
Geological Congress, Volume 2, pp. 69–78
PALAEONTOLOGY

Speaking about ecosystems with respect to the development of sedimentary basin, in this case it will be the Silurian bay (pericontinental sea) of the East Baltic, we are using the first concept since the elucidation of causal connections (mutual relations) or biotic and abiotic components of an ecosystem in the development of the basin is recognized as the main task.

For these purposes we used two models, presenting a generalized description of the basin ecosystem:

- facies-sedimentary model, described in more detail by H. Nestor and R. Einasto (Nestor, Einasto, 1977). It generalizes facies distribution (different types of rocks) and interprets sedimentation conditions;

- ecological model, dealing with a set of more common features of facies-ecological structure of the basin and generalizing relations between organisms and environment (Fig., taken from Kaljo, Vijra, Klaamann et all, 1983).

Actually these are the two sides of one and the same model which could be considered as the model of the basin ecosystem. The development of the basin (of its ecosystem) brought about certain changes in the model resulting from differentiated tectonic movements, eustatics (transgressions - regressions, alternation of facies, inflow of terrigenous material, etc.), climatic changes, organic evolution, etc.

The basin in general

In the Early Palaeozoic the western margin of the East European platform was covered with several marine basins out of which the East Baltic Basin became more distinctly isolated in the Silurian time and retreated to the southwest into the Central European Geosynclinal Basin, probably, of oceanic origin (McKerrow, Ziegler, 1972). The basin was borded by a plain Fenno-Sarmatian continent and was located in a tropic arid zone (Seslavinski, 1978), as a result of which it was subjected to carbonate-fine terrigenous sedimentation. According to the above facies-sedimentary model five main facies belts were distinguished in the basin:

Inshore quiet water (low energy) belt of lagoons and tidal flats, characterized by predominant accumulation of sedimentary dolomitic muds - dolomites and domerites (=dolomitic marls) no with specific poor biota.

Mobile water (high energy) shoal-bar belt, frequent organogenic build-ups and banks. The belt is characterized by the occurrence of variegated grained calcareous sediments (biomorphous, biohermal, detrital, clastic, pelletal limestones) with abundant remains of sessile benthos.

Quiet water belt of open shelf (bottom below the wave base) with admixture of grained and muddy terrigene-carbonaceous sediments (muddy -detritic limestones) with variegated shelly fauna.

Off-shore quiet water transitional belt or gentle slope of the carbonate shelf with the carbonaceous clayey muds (marls and clays, sometimes cryptocrystalline limestones) comprising the remains of some benthic and various pelagic organisms.

Deep stagnant-water belt which at certain stages of development of the basin formed an axial depression of the marginal platform sea. This belt is characterized by the graptolitic clays and argillites, slightly enriched with organic matter, fossil benthic organisms were practically lacking.

Conventionally the above facies belts are referred to as lagoon, shoal, open shelf, slope and depression, accordingly (Fig.).

Ecosystems of the basin

As a result of ecological control each facies belt had its specific set of organisms - an association, which rather frequently differed from that of the neighbour belts, but for a longed time remained stable in its general lines within the boundaries of one facies. In view of the discrete character of biotic and abiotic components of the facies - bathymetric belts we may consider their complex for each belt as an ecosystem of the secondary order of the Silurian East Baltic Basin.

Lagoon ecosystem: Lagoonal rocks have been preserved to our days at some stratigraphic levels only and for that reason this ecosystem has been studied poorly. Restricted circulation of water masses and accompanying deviations in hydrochemical regime were the main ecological factors. The bottom was muddy. Taxonomic composition of the lagoonal biota was rather poor. It was represented by soft-bodied mud-eaters, specific ostracodes (leperditids brachiopods (lingulids) and some bivalve, since the Late Wenlock appeared osteostracans and eurypterids. According to the trophic classification by E. Turpaeva (1954) these organisms are referred to swallowers, collectors, filtrators and active-predators.

Representatives of the first two groups preferred muddy bottom with organic detritus. From autotrophic organisms blue-green algae were widely distributed under lagoonal conditions, creating stromatoliths.

The above association is characteristic for the lagoons with remarkably deviating hydrochemical conditions. Sublagoonal conditions favoured also the development of some specific species of articulate brachiopods, tabulates and stromatoporoids (e.g. *Howelella cuniculi*, *Parastriatopora commutabilis*, *Araneosystroma stelliparratum* in the Rootsikula time of the Late Wenlock).

Shoal ecosystem provided variegated and changable living conditions. In this facies belt the sea bottom consisted of a series of skeletal sand bars alternating with more quiet water depressions. Frequently bioherms complicated the situation. As the sediments of this facies were accumulated under the effect of active waves which sorted and transported the material, the shoal taphocoenosis may considerably differ from the primary biocoenosis. Hydrodynamics served as the first-rate ecological factor, and therefore the shoal ecosystem was dominated by sessile or liberosessile benthos (echinoderms, corals, stromatoporoids, bryozoans, calcareous algae), which rather often formed organic build-ups. Most frequently the liberosessile organisms formed massive or irregular reptant colonies, more resistant to wave action. In shelly fauna (sensu stricto) leading role was plaid by brachiopods. Most typical were strophomenids, often atrypids and rhynchonellids, sometimes large pentamerids, forming bsnks. (e.g. well-known Borealis bank in the Juuru Regional Stage, Llandovery).

Predominantly skeletal sandy sea bottom of the shoal facies belt was not favourable for swallowers and collectors. Therefore members of these trophic groups - mud-eaters, trilobites, ostracodes were rare. Variable filterers and awaiters (corals were more common in this belt. Shoal deposits do not practically contain fossils of planctonic organisms - graptolites and chitinozoans. However, in return, the remains of some nektonic forms are abundant, since the Wenlock in particular, when the conodonts were added to agnathans and fishes. Only in shoal deposits the so-called bone-beds consisting abundant skeletal elements of different vertebrate taxa have been established so far. However, it cannot be acknowledged as a proof indicating that the shoal facies provided optimal conditions for agnathans and fishes, as it might have been purely taphonomic reasons that lead to the formation of bone-beds.

Open shelf ecosystem is characterized by the lack of essential restricting ecological factors. The occurrence of bottom algae shows that the depth of the open shelf sea remained within the euphotic zone. The bottom consisted of terrigenous carbonate muds mixed with shell detritus. Such a bottom provided favourable conditions for the development of organisms inhabiting muddy bottoms as well as those preferring had substrate. Owing to the above conditions the open shelf biota was taxonomically rather rich, however, without distinct domination of any group. From benthic organisms brachiopods appeared to be most abundant and variegated. Conodonts prevailed among pelagic groups. Open shelf deposits were rather rich in chitinozoans whereas graptolites were quite rare, mostly specific epipelagic forms (e.g. Climacograptus, Paraclimacograptus, etc., see Kaljo, 1978). The fauna was richer in this part of the open shelf which was bordering with the shoal belt. There occurred also colonial corals, stromatoporoids, clacareous algae with their quantity decreasing gradually towards the open sea.

At some stages of the development of the Baltic Silurian Basin the open shelf ecosystem experienced some changes. In the Early and Middle Llandovery calcareous muds accumulated on the shelf from time to time. They were remarkably poorer in biota consisting mainly of single brachiopods (e.g. Borealis pumilus), small rugose corals (Densiphyllum) and abundant dendroids (Leveillites). Such a situation was due to inconsiderable inflow of terrigenous material, however, it might have been associated with some hydrochemical changes in the basin as well.

In the Late Silurian at the final stage of the development of the East Baltic Silurian Basin the assumulation of pelitic terrigenous material became more intensive (marls with limestone interlayers were very common). Quantitative correlations of different groups underwent some changes, in comparison with Early Silurian the share of ostracodes and bryozoans increased, that of tabulates, stromatoporoids, in particular decreased.

Slope ecosystem. Marls of this facies belt show signs of quiet water sedimentation, evidently, already below the euphotic zone. The latter is proved by the lack of bottom vegetation. In this belt benthic organisms were not so frequent and variable as those on the open shelf. Brachiopods, ostracodes and trilobites in particular, dominated being represented, as a rule, by other communities rather than in shelf ecosystems (see Kaljo, Klaamann, eds, 1982). Rugose corals were still rather frequent in the upper part of the slope. Single tabulates (halysitids and heliolitids), and rare stromatoporoids (densastromatids) occurred. The share of trilobites increased with the depth. Slope facies rocks bore some tracks and trails of mud-eaters, however, they were less frequent than in lagoonal and open shelf deposits. In this thus, in the slope ecosystem collectors became dominants among the trophic groups, whereas filterers and swallowers lost them considerably.

Among the pelagic organisms chitinozoans were of the widest distribution in the slope facies belt. Conodonts, agnathans and fishes were a bit less frequent here than in the open shelf belt, whereas the number of graptolites increased toward the deeper water facies.

Depression ecosystems. Remains of benthic organisms are practically lacking in clayey sediments of this part of the basin, however, they do contain abundant remains of pelagic graptolites, accompanied by hyolithids, nautiloids and chitinozoans. Conodonts as well as the agnatha- and fish-remains are nearly lacking. Graptolitic clays contain 7-18% of organic matter, which together with the absence of bottom fauna refers to the stagnant water anaerobic conditions in the near-bottom layers of water of the central deep part of the basin.

Depression facies appeared to be absent in the East Baltic territory at the beginning (Early Llandovery) and at the end of the Silurian (Late Ludlow - Pridoli), when this facies remained behind the boundaries of the region.

Some conclusions

A full description of a palaeoecosystem of the basin could be obtained only through complex investigation of both the rocks and fossil organisms. However, in this respect, a considerable number of sections and palaeoecosystems of the East Baltic Silurian have been insufficiently studied so far. On the other hand, we cannot set up a task to study all the components of the system equally completely. It would sound unreal. That will do, if to concentrate attention on the most characteristic features of lithology and biota, enabling to analyse and interpret a palaeoecosystem. The results obtained through preliminary studies of different character might provide basis for the selection of the components to be subjected to detailed studies.

Our results on the East Baltic indicate close connection between organisms and environment revealed by the specifics of communities and associations of different stages of the basin development. However, much is still to be done for the elucidation of the details of mutual relations, in order to make wider and more precise use of the fossil organisms as indicators of the environment.

REFERENCES

Kaljo, D.L. On the bathymetric distribution of graptolites.- Acta Palaeont. Polonica, 1978, vol. 23, No. 4, p.523-531.

Kaljo, D.L. Ecostratigraphy - some aspects from East Baltic practice.- In: Ecostratigraphy of the East Baltic Silurian. "Valgus", Tallinn, 1982, p.9-16.

Kaljo D.L., Klaamann E.R. Communities and biozones of the Baltic siliman. "Valgus", Tallin, 1982, p.138.

Kaljo D.L., Viira V.Ja., Klaamann E.R., Miannil R.P., Miarss T.I., Nestor V.V., Nestor Kh.E., Rubel M.P., Sarv L.I., Einasto P.E. The ecological model of

the East Baltic silurian basin.- In the book "Problems concerned with the ecology of the flora and the fauna of the oldest basins". M., Nauka, 1983.
Krasilov V.A. Paleoecosystems - "Izvestia of the USSR Academy of Sciences, ser. geol., 1970, N 4, p.144-150.
McKerrow W.S., Zeigler, A.M. Paleozoic oceans.- Nature (Phys. Sci.), 1972, vol. 240, p.92-94.
Nestor H.E., Einasto P.E. Facial sedimentological model of the Silurian Baltic pericontinental basin. In the book "Facies and the silurian Baltic fauna. Tallin, the USSR Academy of Sciences, 1977, p.89-121.
Seslavinskii K.B. Climate and global climatic belts of Ordovician and the Silurian Global belts.- "Izvestia of the USSR Academy of Sciences, ser. geol., 1978, N 3, p.98-112.
Sokolov B.S. Major ways of complex geological studies. "Izvestia of the USSR Academy of Sciences, ser. geol., 1978, N 12, p.5-16.
Turpaeva E.P. Types of marine bottom biocoenoses and the dependance of their distribution of abiotic factors of environment.- "Proceedings of the oceanological Institute, 1954, v. XI, p.36-55.

Fig. 1. Ecological model of the Silurian East Baltic Basin presenting composition and distribution of main benthic and pelagic associations in differnet ecosystems. Twice as large conventional signs mark the group in this ecosystem where its maximum occurs. The lower half of the figure lists main trophic groups of benthos and the character of the substrate in a given ecosystem.

Conventional signs: 1 - stromatoporoids; 2 - massive tabulates; 3 - single rugose corals; 4 - tabulates; 5 - brachiopods; 6 - bivalves; 7 - trilobites; 8 - ostracodes; 9 - eurypterids; 10 - crinoids; 11 - graptolites; 12 - chitinozoans; 13 - scolecodonts; 14 - conodonts; 15 - agnathans and fishes; 16 - stromatoliths; 17 - tracks of mud-eaters.

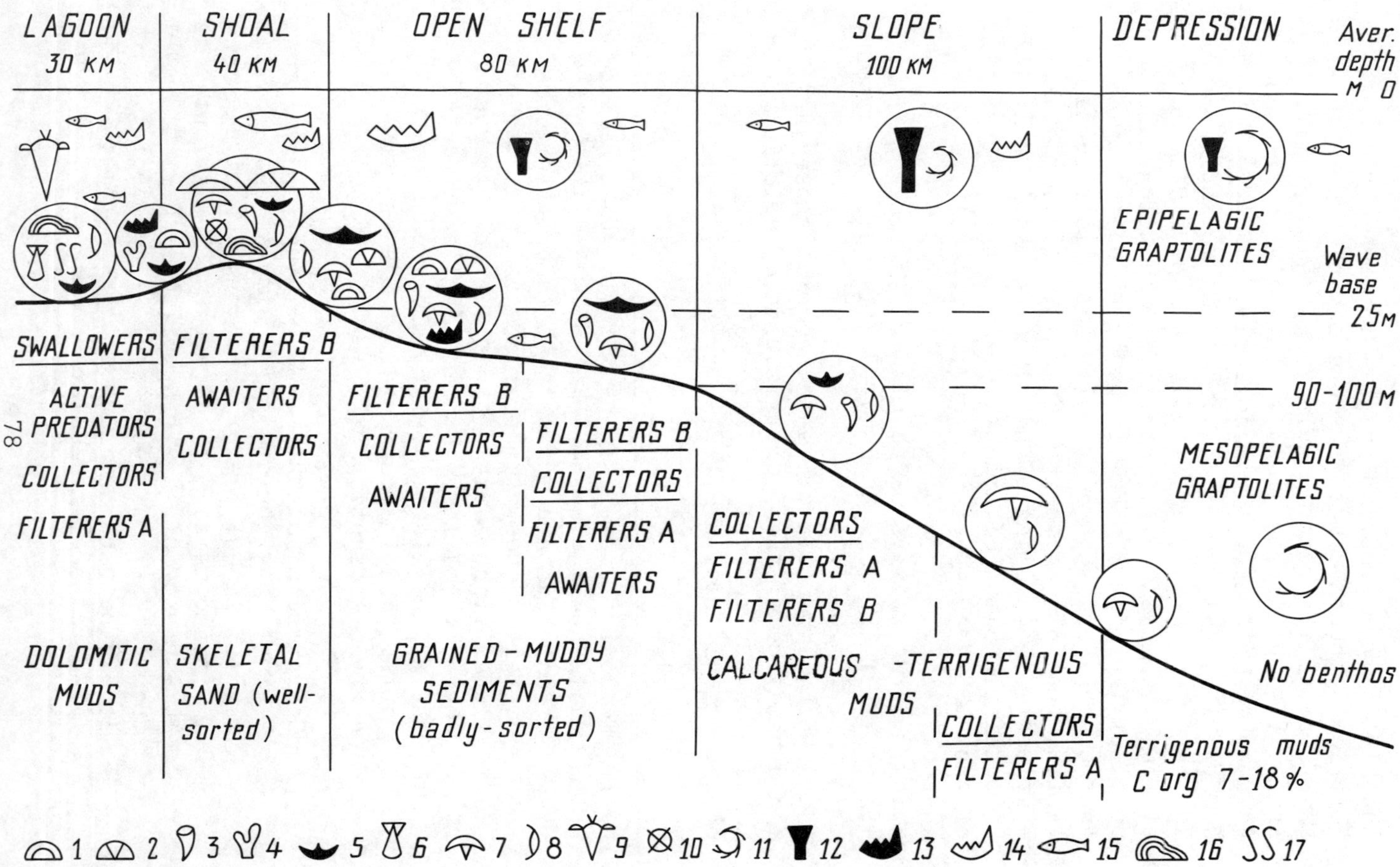

LAGOON
30 KM
SHOAL
40 KM
OPEN SHELF
80 KM
SLOPE
100 KM
DEPRESSION
Aver. depth M 0
EPIPELAGIC GRAPTOLITES
Wave base
25M
90-100M
MESOPELAGIC GRAPTOLITES
SWALLOWERS
ACTIVE PREDATORS
COLLECTORS
FILTERERS A
FILTERERS B
AWAITERS
COLLECTORS
FILTERERS B
COLLECTORS
AWAITERS
FILTERERS B
COLLECTORS
FILTERERS A
AWAITERS
COLLECTORS
FILTERERS A
FILTERERS B
COLLECTORS
FILTERERS A
No benthos
DOLOMITIC MUDS
SKELETAL SAND (well-sorted)
GRAINED-MUDDY SEDIMENTS (badly-sorted)
CALCAREOUS -TERRIGENOUS MUDS
Terrigenous muds C org 7-18 %
1 2 3 4 5 6 7 8 9 10 11 12 13 14 15 16 17

BIOMINERALIZATION PROCESSES AND PRODUCTS AND THE EVOLUTION OF BIOMINERALIZATION

HEINZ A. LOWENSTAM
Division of Geological & Planetary Sciences
California Institute of Technology
Pasadena, California 91125, U.S.A.

ABSTRACT

This review summarizes briefly first the present status of our knowledge of the diversity of minerals formed by extant organisms, the phyla involved in biomineralization, the physical and chemical properties which distinguish these minerals from their inorganic counterparts and the processes leading to their formation.

The succeeding section dealing with the history of biomineralization is designed to show that the information, now at hand, on extant mineral forming agents, mineralization processes and their products seems to present a different perspective on the early phase of the evolution of biomineralization than has been usually inferred from records of body fossils. Sedimentary records of individual mineral grains and minute mineral aggregates of biologic origin are considered likely to resolve the controversy as well as to elucidate in greater detail the history of biomineralization.

Proceedings of the 27th International
Geological Congress, Volume 2, pp. 79–95
PALAEONTOLOGY

MINERAL DIVERSITY

Forty different minerals are by now known to be formed non-pathologically by extant organisms (Lowenstam and Weiner, 1983), Table 1 lists the minerals and shows their known distribution in prokaryotes and eukaryotes. Three overall trends recognized by Lowenstam (1980), when only 31 biogenic minerals were known, are still applicable, even though two of them have been proportionately somewhat modified. These trends adjusted for the new data, show that of the minerals listed in Table 1, about one-half are Ca^{++} minerals, one-half are H_2O and OH-containing minerals and one-fourth are colloidal materials. The distribution of the mineral species within the phyla spectrum, as shown in Table 1, adds new evidence in support of the long established fact that carbonate minerals are by far the most widely utilized minerals by extant organisms. Opal has emerged as the second most extensively utilized biologic mineral. Ferrihydrate and related ferric oxide minerals rank third. Magnetite, considered as a biological oddity when first discovered as a biomineralization product in chiton teeth (Lowenstam, 1962), is now known to be synthesized widely by organisms ranging from prokaryotes to eukaryotes (Kirschvink, 1980). It may well prove to be the fourth most extensively formed biosynthesis product (Lowenstam, 1981).

MINERAL FORMING PROCESSES

Two distinct mineralization processes have been recognized (Lowenstam, 1981). One of these has been termed "biologically induced" mineralization. In this process minerals are formed by interaction of biologically produced

Table 1. The diversity and distribution of biogenic minerals in extant organisms.
The term "precursor" refers to an amorphous phase which upon heating to 500°C converted to the designated crystalline form.

	MINERALS	MAJOR CATIONS	Monera	Dinoflagellata	Haptophyta	Bacillariophyta	Phaeophyta	Rhodophyta	Chlorophyta	Zygnematophyta	Chrysophyta	Rhizopodea	Siphonophyta	Charophyta	Heliozoata	Radiolariata	Foraminifera	Mixomycota	Ciliophora	Basidiomycota	Deuteramycota	Mycophycophyta	Bryophyta	Tracheophyta	Porifera	Cnidaria	Platyhelminthes	Ectoprocta	Brachiopoda	Annelida	Mollusca	Arthropoda	Sipuncula	Echinodermata	Chordata
			MONERA	PROTOCTISTA																FUNGAE			PLANTAE		ANIMALIA										
	Calcite	Ca	O	O	O			O					O	O			O	O					O	O	O	O	O	O	O	O	O	O	O	O	O
	Aragonite	Ca	O		?		O	O	O				O				O							O	O	O		O		O	O	O	O		O
	Vaterite	Ca						O																O						O	O	O			O
CARBONATES	Monohydrocalcite	Ca	O																												O				O
	Protodolomite	Ca,Mg																																O	
	Amorph hydrous carbonate	Ca																O								O	O				O	O			O
	Francolite	Ca																											O		O				O
	Dahllite	Ca	O																O		O			?											O
	$Ca_3Mg_3(PO_4)_4$	Ca,Mg																												O					
	Huntite	Mg,Ca																																	O
PHOSPHATES	Brushite	Ca																													O				O
	Octacalcium phosphate	Ca																																	O
	Calcium pyrophosphate	Ca																													O				
	Amorph dahllite precursor	Ca																										O			O	O			O
	Amorph brushite precursor	Ca																													O				
	Amorph. whitlockite precursor	Ca																									O			O	O				
	Amorph. - Fe - Ca - phosphate	Fe, Ca																												O	O			O	
HALIDES	Fluorite	Ca																													O	O			
	Amorph fluorite precursor	Ca																													O				O
	Whewellite	Ca							?	O			O	?						O	?	O	O	O								O			
OXALATES	Weddelite	Ca																		O	?			O							O			O	O
	Glushinskite	Mg																				O													
CITRATES	Calcium citrate	Ca																														O			
	Gypsum	Ca							O	O													?	O		O									
SULFATES	Celestite	Sr														O																			
	Barite	Ba							O			O		O																					
SILICA	Opal	Si		O		O	?				O	O			O	O	O							O	O					O	O	O		O	
	Magnetite	Fe	O						?																						O	O			O
	Goethite	Fe																													O				O
Fe-OXIDES	Lepidocrocite	Fe	?																						O						O				
	Ferrihydrite	Fe	O																	O	O		O	O						O	O				O
	Amorph. "ferrihydrites"	Fe	O														O													O					O
	"Amorph. ilmenite"	Fe,Ti																																O	
Mn-OXIDES	Todorokite	Mn	O																																
	Birnessite	Mn	O																																
	Pyrite	Fe	O																																
	Hydrotroilite	Fe	O																																
SULFIDES	Sphalerite	Zn	O																																
	Wurtzite	Zn	O																																
	Galena		O																																

PHYLA.

metabolite end products with cations in the external environment. The other process has been termed "organic matrix mediated" mineralization. In this process nucleation and subsequent development of minerals takes place in contact with a preformed organic matrix. The two processes constitute end members of a spectrum in which the organism exercises increasing control over the kind of minerals to be produced as well as mineral growth. In "biologically induced" mineralization processes which denote the spectral end member of least biologic control, the mineral precipitates resemble those formed by inorganic processes in crystal habits. Crystal aggregates are devoid of unique shapes and usually show random orientation of variously sized crystallites. There is a tendency for less vigorous control of these processes on the mineral phases of the precipitates formed. The $^{34}S/^{32}S$ ratios of pyrites precipitated by bacterially induced sulfate reduction shows, unlike inorganically formed pyrites, disimiliatory fractionation of the sulfur isotopes (Thode, et al., 1951). At the other end of the spectrum in "matrix mediated" mineralization processes, as a rule, the minerals have unique crystal habits and a narrow size distribution. Polycrystalline bodies adopt genetically controlled morphologies and show well ordered crystal axes orientation as well as crystal fabrics (Lowenstam, 1980; Lowenstam, 1981; Weiner and Traub, 1980). Several of the minerals formed today by "matrix mediated" processes in the hydrosphere are not known to be precipitated there by inorganic processes. Examples are the minerals magnetite, fluorite, gypsum and celestite (Lowenstam and Weiner, 1983). Minerals formed by these processes tend to have an overprint of disequilibrium chemical signatures with reference to the environment

(Lowenstam, 1981). At present this is best documented by the Sr and Mg contents and the oxygen and carbon isotopic composition of carbonate minerals (summarized in Milliman, 1974; Dodd and Stanton, 1981).

Biochemical and x-ray diffraction studies of the organic matrix, based on mollusc shells have shown that the matrix consists of a core component composed of a silk-fibroin-like protein and commonly also of β chitin (Weiner and Traub, 1980; Weiner and Traub,, 1981). The core component is veneered on both sides by acidic surface layers. A proposed model envisions that the function of the acidic surface layers include mineral nucleation, crystallographic axes orientation, as well as crystal growth inhibition. The core component is considered to provide a substrate for the acidic proteins and performs mechanical functions. A review by Weiner, et al., (1983) elaborates on the matrix structure and function of molluscs and other eukaryotes.

"Biologically induced" mineralization processes occur widely among prokaryotes, intra-and extra-cellularly. Among the eukaryotes the mineralization products are known to occur in some algal representatives of the proterozoic, in many plant phyla and apparently also in many animal phyla. The minerals form in the algae inter-cellularly and in the plants and animals intra-cellularly (Arnott, 1973; Borowitzka et al., 1974; Lowenstam, 1981). Assignment to "biologically induced" mineralization processes is in most instances based by this author on the crystal habits, shapes and fabrics of aggregates as seen in published illustrations of biominerals.

The "organic matrix" mediated mineralization processes are widespread among the eukaryotes. The mineralization products are as a rule located at extracellular sites where they form commonly mineralized hard parts such as e.g. exo-and endo-skeletons but they may consist solely of numerous isolated mineral grains or skeletal pinpoint mineralization by crystal aggregates (Lowenstam and Margulis, 1980).

Recent reports on bacterially formed minerals with unique crystal habits and in two cases with well ordered crystal fabrics have been interpreted to indicate that prokaryotes are also involved in "matrix mediated" mineralization processes (Lowenstam and Weiner, 1983). The observations in support of this conclusion were as follows. Magnetotactic bacteria discovered by Blakemore (1975) form intracellularly, membrane bound magnetite crystals with hexagonal prism shape (Towe and Moench, 1981; Matsuda et al., 1983), or "tear drop" shape (Blakemore et al., 1981). The crystals have a narrow size distribution (Towe and Moench, 1981). In addition, an unnamed species of _Leptotrix_ has been shown to form a hexagonally shaped grill work of an as yet unidentified ferric-oxide mineral inside the sheath which covers the cell filaments (Caldwell and Caldwell, 1980). The unique crystal fabric calls for organic framework control, but it is not known how the enclosing sheath serves as the organic matrix. Other examples are provided by two cyanobacterial species of _Geitleria_. The filament surfaces are thickly covered by an orderly stacked fabric of calcite crystals which form a square shaped grill work (Friedmann, 1955, 1979). Treatment of the calcite coat with dilute hydrochloric acid leaves a delicate "pectinous" sheath as a residue (Friedmann, 1955; Lowenstam and Weiner, 1983). SEM micrographs of critical point dried and EDTA decalcified samples of freshly collected _Geitleria_ specimens have been recently obtained by the writer in his current research. The micrographs show that the spaces bordered by the square shaped frames of the calcite crystals are the sites of organic matrices which are interconnected to form a continuous framework within the mineralized coat of the bacteria.

This observation extends definitive documentation of "organic matrix mediated" mineralization among prokaryotes from the magnetotactic bacteria to the cyanobacteria and further shows that these processes may take place extra-cellularly as well as intra-cellularly. The mineralized structures formed by *Geitleria* on the filament surfaces grossly resemble skeletal structures of eukaryotes with extensive matrix development but seem to differ from them in their crystal arrangement.

"Biologically induced" and "matrix mediated" processes have been occasionally noted to occur in one organism at different anatomical sites and even at the same mineralization site (Lowenstam and Weiner, 1983). Known examples are at present limited to animal representatives of the eukaryotes.

HISTORY OF BIOMINERALIZATION PROCESSES: FOSSIL AND RECENT PERSPECTIVES

Until recently, information on the history of biomineralization was derived almost entirely from body fossils, in the form of mineralized skeletal remains. The use of $^{34}S/^{32}S$ ratios as a tracer of "biologically induced" mineralization by sulfate reducing bacteria in Precambrian sedimentary pyrites has demonstrated that geochemical approaches can contribute materially to the elucidation of the history of biomineralization. It appears that "biologically induced" type mineralization was the first biomineralization mode to evolve in the Precambrian. This supposition is based on the sulphur isotopes from sedimentary pyrites which show the beginning of dissimilatory fractionation by sulfate reducing bacteria in samples of 2 to 2.4 billion years in age (Cameron, 1982; Skyring and Donnelly, 1982). Only "bacterially induced" mineralization is in evidence up to near the end of the Proterozoic. Beginning with the Vendian - early Cambrian transition eukaryotes began to exercise more rigorous control over mineral forming processes and "organic matrix mediated" mineralized evolved. As shown by skeletal remains, animals were the first eukaryotes to develop "matrix mediated" mineralization. Once initiated, this mineralization process began to spread rapidly to a wide range of animals in early Cambrian time and continued to proliferate among the eukaryotes in the course of the Phanerozoic.

In extant prokaryotes and eukaryotes there are examples of mineralization processes which cover the spectrum between "biologically-induced" and "matrix mediated" mineralization. The fossil record appears to present a very different perspective of the history of biomineralization, namely, that "matrix mediated" mineralization was from the beginning limited to the eukaryotes and evolved considerably later in geologic time than "biologically induced" mineralization which continued to be utilized only by prokaryotes. A number of seemingly unrelated aspects of present day "matrix mediated" mineralization products have raised serious doubt about the validity of the inference from the fossil records that mineralized hard parts which began to appear first in the geologic record at the Vendian - Cambrian transition actually mark the inception of "matrix mediated" processes and that animal representatives of the eukaryotes were the initiators of these processes.

It has been difficult to conceive that from its inception "matrix mediated" mineralization should have been perfected as a process to yield in animals uniformly and sufficiently massive skeletal mineralization to preserve their products morphologically intact. Based on the observations that a number of early Cambrian skeletal records have only weak surficial mineralization, it had been postulated that pinpoint mineralization may have preceded massive mineralized hard parts (Lowenstam and Margulis, 1980). The indications among representatives of the late Vendian soft bodied Ediacara fauna of partial skeletal mineralization and, the development of spicules in pennatulid-like forms (Glaessner and Daily, 1959; Glaessner, 1976) seem to support the notation of an initial phase of patchy skeletal mineralization.

Similar trends are in evidence with respect to increase in complexity in organization of mineral products that are formed at homologous sites in several extant animal groups, as e.g. the molluscs and fishes, (Lowenstam, 1980; Lowenstam and Weiner, 1983). Reflecting evolutionary trends, they seem not only to lend support to the postulate that pinpoint mineralization preceded total skeletal mineralization in animals, but seem to provide, further, a clearer concept of the consecutive stages that trace the pathways of these processes as shown in the following. Stage 1 is characterized by single crystals or single colloidal bodies in great profusion at the depositional site. At Stage 2, the constituents consist of a profusion of minute separate crystal aggregates instead of single crystals. At Stage 3, most of the crystals are incorporated into a larger single body, but a few minute crystal aggregates remain isolated. Stage 4, marks the aggregation of all crystals into a single structure without differentiation in crystal fabric and mineralogy. Stage 5 resembles the previous stage by the aggregation of all crystals into a single body but differs from Stage 4 in the subdivision of the structures into discrete microarchitectural units with different crystal fabrics which do not have to be mineralogically distinct. In some groups of organisms the trend terminates at Stage 4 and in others, one or another intermediate stage may be missing. As is usual for "matrix mediated" mineralization products, the crystals formed at Stage 1 have commonly specific crystal habits, may be unique in mineralogy, and commonly have an overprint of disequilibrium chemical signature with reference to the environment. From the beginning of localized crystal aggregation into multiple bodies at Stage 2 to the incorporation of all crystals into single structures, as formed at Stages 4 and 5, one finds well ordered crystal fabrics and well defined, genetically controlled morphologies of the polycrystalline bodies. Hence, individual crystals produced at Stage 1 and minute crystal aggregates formed at Stages 2 and 3 should be distinguishable from the crystals and crystal aggregates of "biologically induced" and inorganic origin.

The sum of the considerations outlined above presents a very strong argument that at the time the first completely mineralized animal skeletons made their appearance in the fossil record, "organic matrix mediated" processes had not just come into existence, but had already evolved in evolutionary terms to near completion, that is, definitely to Stage 4 and possibly Stage 5. Therefore, "matrix mediated" mineralization should have come into existence earlier in Protozoic time and the seemingly unrecorded evolutionary initial stages in development by these processes should be traceable in the sedimentary rocks by means of single crystals and minute mineral aggregates, characterized by the criteria outlined above. A further implication of the possibility that the early evolution of "matrix mediated" processes took place in Proterozoic time, is, that animal representatives may not have been the initiator of these processes, contrary to the current view which has been based on body fossils.

Independently derived support for a Precambrian origin of "matrix mediated" mineralization processes may be seen in an inter-phylum survey of the biochemistry of organic matrices from the mineralized hard parts of eukaryotes (Weiner et al., 1983). The data comes from a wide range of animal phyla and one group of proterozoic. This survey reveals common interphyla biochemical properties of the eukaryotic organic matrices, an observation which is consistent with the notion that "matrix-mediated" mineralization "know how" was inherited from some common Precambrian ancestral stock. The widespread utilization of matrix mediated mineralization for constructing solid exoskeletons that came into existence at the end of the Precambrian would then again imply the culmination of a long evolutionary history and not the inception of this mode of biomineralization. The alternative that each phylum "invented" independently the matrix strategy would have to be qualified by invoking an explanation for the observed common inter-phylum biochemical properties of organic matrices.

As noted above, it has been recently shown that extant prokaryotes also produce minerals with the aid of organic matrices. One of them, the magnetotactic bacteria have magnetite crystals with unique crystal habits and are characterized by a narrow size distribution. Magnetite crystals of this type have been recently located in deposits of Miocene age (Kirschvink, 1982; Chang and Kirschvink, 1983), but there are as yet no data on the geologic range of these magnetite forming bacteria nor on the two other mineral forming bacterial groups. It is significant, however, that the extant magnetite and ferric oxide producing bacteria may still reflect the reducing condition and change over from reducing to oxidizing conditions in the biosphere during the early Precambrian. It is of further interest that the third group of mineral forming bacteria belongs to the cyanobacteria which extend in the sedimentary record back to 3.5 years B.P. (Awramic et al., 1983). Nothing is as yet known about the biochemistry of the organic matrices from prokaryotic mineralization products and hence we do not as know as yet whether they are analogous to eukaryotic matrices. Clarification of this aspect and also a search for evidence of the existence of Precambrian prokaryotes involved in "matrix mediated" mineralization is needed. Given this information it should be possible to ascertain whether "matrix mediated" mineralization processes evolved independently in prokaryotes and eukaryotes, or whether the eukaryotes inherited the "know how" of these processes from a prokaryote progenitor in the Precambrian.

It is of interest to note that an inheritance of the "know-how" of "matrix mediated" mineralization by eukaryotes from prokaryotes can be readily accommodated in the evolutionary scheme visualized by Margulis (1981) since in this scheme the eukaryotes evolved directly from the prokaryotes, whereas in the scheme of Woese (1981) the origin of the "know-how" is as yet not as clearly discernable.

REFERENCES

Arnott, H.J. (1973). Plant Calcification. In: Biological Mineralization, I. Zipkin, (ed) John Wiley and Sons, New York, pp. 609-627.

Awramik, S.M. Schopf, J.A. and Walter, M.R. (1983). Filamentous fossil bacteria from the Archean of Western Australia. Precambrian Research. 20, 357-374.

Blakemore, R.P. (1975). Magnetotactic bacteria. Science 190, 377-379.

Blakemore, R.P., Frankel, R.B. and Kalmijn, A.J. (1981). South-seeking magnetotactic bacteria in the Southern Hemisphere. Nature. 286, 384-385.

Borowitzka, M.A., Larkum, A.W.D. and Nockolds, C.E. (1974). A scanning electron microscope study of the structure and organization of the calcium carbonate deposits of algae. Phycologia, 13, 195-203.

Caldwell, D.E. and Caldwell, S.J. (1980). Fine structure of in situ microbial iron deposits. Geomicrobiol. Jour. 2, 39-53.

Cameron, E.M. (1982). Sulphate and sulfate reduction in early Precambrian oceans. Nature, 296, 145-148.

Chang, S.R. and Kirschvink, J.L. (1983). Single-domain magnetite particles in deep-sea sediments: Possible bacterial microfossils. Geol. Soc. Amer. Abstr. Program, 15, 542.

Dodd, J.R. and Stanton, R.J. (1981). Paleoecology, Concepts and Applications. John Wiley and Sons, New York, pp. 138-142.

Friedmann, I. (1955). Geitleria calcarea n.gen. et n.sp. A new atmophylic lime-encrusting blue-green algae. Botan. Notis. Lund, 108, 439-445.

Friedmann, I. (1979). The genus Geitleria (Cyanophyceae or Cyanobacteria): distribution of G. calcarea and G. floridana n.sp. Pl. Syst. Evol., 131, 169-178.

Glaessner, M.F. and Daily, B. (1959). The geology and late Precambrian fauna of the Ediacara fossil reserve, Rec. South Australian Mus. (Adelaide). 13, 369-401.

Glaessner, M.F. (1976). Early Phanerozoic annelid worms and their geological and biological significance. Jour. Geol. Soc. London, 132, 259-275.

Kirschvink, J.L. (1980). Biogenic magnetite (Fe_3O_4): A ferro-magnetic mineral in bacteria, animals and man. In: Ferrites: Proc. Internat. Conf. Sept-Oct, 1980, Japan, pp.135-138.

Kirschvink, J.L. (1982). Paleomagnetic evidence for fossil biogenic magnetite in western Crete. Earth Planet, Sci. Lett., 59,, 388-392.

Lowenstam, H.A. (1962). Magnetite in denticle capping in recent chitons (Polyplacophora). Bull. Geol. Soc. Amer., 73, 435-438.

Lowenstam, H.A. (1980). Bioinorganic constituents of hard parts. In: Biogeochemistry of Amino Acids, P.E. Hare, T.D. Hoering and K. King, Jr. (eds). John Wiley and Sons, New York, pp.3-16.

Lowenstam, H.A. (1981). Minerals formed by Organisms. Science, 211, 1126-1131.

Lowenstam, H.A. and Margulis, L. (1980). Evolutionary prerequisites for early Phanerozoic calcareous skeletons. BioSyst. 12, 27-41.

Lowenstam, H.A. and Weiner, S. (1983). Mineralization by organisms, and the evolution of biomineralization. In: Biomineralization and Biological Metal Accumulation, P. Westbroek and E.W. DeJong (eds), D.Reidel Publ. Co., Dordrecht, pp.191-203.

Margulis, L. (1981). Symbiosis in Cell Evolution. W.H. Freeman and Co., San Francisco.

Matsuda, T., Endo, J., Osakabe, N., Tonomura, A. and Arii, T. (1983). Morphology and structure of biogenic magnetite particles. Nature, 302, 411-412.

Milliman, J.D. (1974). Marine Carbonates. Springer-Verlag, New York, pp.52-147.

Skyring, G.W. and Donnelly, T.H. (1982). Precambrian sulfur isotopes and a possible role for sulfite in the evolution of biological sulfate reduction. Precambrian Res., 17, 41-61.

Towe, K.M. and Moench, T.T. (1981). Electron-optical characterization of bacterial magnetite. Earth Planet. Sci. Lett., 52, 213-220.

Thode, H.G., Kleerekoper, H. and McElcheran, O. (1951). Isotopic fractionation in the bacterial reduction of sulfate. Research, 4, 581.

Weiner, S. and Traub, W. (1980). X-ray diffraction study of the insoluble organic matrix of mollusk shells. FEBS Lett. III, 312-316

Weiner, S. and Traub, W. (1981). Organic-matrix-mineral relationships in mollusk shell nacreous layers. In: Structural Aspects of Recognition and Assembly in Biological Macromolecules. M. Balaban, J.L. Sussman, W. Traub and A. Yonath (eds). Balaban ISS., Rehovot, pp.467-482.

Weiner, S., Traub, W. and Lowenstam, H.A. (1983). Organic matrix in calcified exoskeletons. In: Biomineralization and Biological Metal Accumulation, P. Westbroek and E.W. DeJong (eds), D. Reidel Publ. Co., Dordrecht, pp.205-224.

Woese, C.R. (1981). Archaebacteria. Scientific Amer., 244, 98-122.

Contribution No. 4024, Division of Geological and Planetary Sciences, California Institute of Technology, Pasadena, CA 91125, U.S.A.

PHYLOGENY OF HIGHER PLANTS AND FLOROGENESIS

MEYEN, S.V.
Geological Institute of the Academy of Sciences
of the USSR, Moscow, USSR.

The study of florogenesis crowns palaeobotany, because it brings together all its main branches. Florogenetics includes analysis of (1) the phylogenetic relations between plant taxa, (2) the dispersal of taxa, and (3) the history of palaeofloristic units (palaeophytochoria) against a background of environments.

Special florogenetic publications based on palaeobotanical material are not numerous. The best known are a monograph of Seward /11/ and works of Kryshtofovich /5, 6/. Florogenetic chapters are included in some text-books /4, 15/. There are a number of palaeobotanical papers treating separate aspects of florogenetics, e.g. the paper of Schopf /10/ summarizing relations between Gondwana and Northern floras with reference to the phylogeny of major taxa of higher plants.

The present paper attempts to summarize the achievements of both the phylogenetics of higher plants and palaeofloristics during last decades under the florogenetic viewpoint. The aim of the paper is to outline main features of florogenesis during the Palaeozoic and, partly, Mesozoic. The palaeofloristic units adopted below (Fig. 1) have been repeatedly described in the literature /1, 12, 13/. The mobilistic views are shared by the author.

The present author has compiled a phylogenetic tree of higher plants (Fig. 2) showing only higher taxa. Beginning from the Carboniferous upwards, for each taxon its geographical distribution is shown

Proceedings of the 27th International
Geological Congress, Volume 2, pp. 97–109
PALAEONTOLOGY

according to three major biotas - equatorial belt (EB) with adjacent ecotone areas, and two extraequatorial biotas. The distribution of these biotas in the Late Palaeozoic is shown on an inset of the scheme. In the Triassic, roughly the same pattern of phytochoria persisted. In the Jurassic and Early Cretaceous, a southern counterpart of the Boreal extraequatorial biota cannot be reliably recognized. Beginning with the Jurassic the shading on phylogenetic branches corresponding to the extraequatorial biota is shown conventionally. This shading tentatively denotes the distribution of pertinent taxa in those parts of Gondwana which are more remote from the Equator. The Devonian part of the phylogenetic tree yields no geographic distribution of taxa, because recognizable phytochoria of the period are yet to be established /3/. It should be mentioned, however, that the Devonian phylogeny of higher plants was reconstructed essentially on the basis of West European and North American materials. The phylogenetic lineages shown on the scheme are explained below.

Lycopsids are shown on the scheme according to remains of all kinds, including dispersed megaspores. The latter are much more common in the fossil record than megafossils. Megaspores found in the absence of pertinent megafossils might have belonged to herbaceous forms (e.g. selaginellas). If one takes into account only megafossils, the distribution of the group would be punctuated. For instance, lycopsid megafossils are unknown in the Permian of the inner parts of Angaraland (Angara area), though megaspores occur intermittently. In both Gondwanaland and Angaraland the appearance of lycopsids within the Middle Carboniferous - Upper Permian interval probably marks warming episodes. In both the continents, forms similar in the structure of their vegetative parts of genera of the Upper Devonian - lowermost Carboniferous long persisted.

The origin of the Bowmanitales (sphenophylls) is unclear. Commonly they are clumped together with the remaining articulates all being regarded as derived from cladoxylaleans. However, the independent origin of bowmaniteans cannot be excluded. They differ from other articulates (Calamostachyales, Equisetales) in a number of essential characters, viz. exarch or slightly mesarch protostele, and adaxial sporangiophores. In these two characters bowmanitaleans are closer to lycopsids. The genera like <u>Eleutherophyllum</u> and <u>Estinnophyton</u> occupy an intermediate position between Bowmanitales and lycopsids. In the meantime, on the scheme the Bowmanitales are not derived from lycopods, because such decision would be purely hypothetical. Another plant group from which the Bowmanitales might have been derived is the order Ibykales (= Iridopteridales). This order could also be ancestral to the Cladoxylales. In Ibykales the stele is similar to that of the Bowmanitales, and the verticillate arrangement of lateral appendages is recorded.

The evolution of Bowmanitales proceeded chiefly in EB. They occur elsewhere in separate horizons probably marking episodes of warming. The youngest members of the order occur sporadically in the Permotriassic of the Siberian platform.

The orders Calamostachyales (= Calamitales) and Equisetales appear first in EB, and soon (even in the Early Carboniferous) in the extraequatorial biotas. In the latter regions very primitive types persisted, belonging to the families Tchernoviaceae (Angaraland) and Gondwanostachyaceae (Gondwanaland) until the Permotriassic. These two families are comparable in both fructification structure and vascularization of axes to the Early Carboniferous <u>Archaeocalamites</u> - like plants of EB.

Ferns (Polypodiopsida; orders Zygopteridales and Polypodiales s.l.) are sufficiently studied only in the EB. In the extraequatorial Upper Palaeozoic biotas the ferns are much less frequent and their affinities are rarely known. Primitive ferns (Geperapteris, Prynadaeopteris) similar to Carboniferous equatorial zygopteridaleans occur in Angaraland until the Late Permian. Most of the discoveries of Permian osmundaceous ferns have been confined to Angaraland and Gondwanaland. The main divergency of ferns seems to have proceeded in EB.

The Carboniferous progymnosperm orders Protopityales and Archaeopteridales occur rarely in EB only. The genus Fedekurtzia of the pre-Glossopteris Gondwana flora can be affiliated with progymnosperms. The orders Noeggerathiales and Tingiales recorded only in EB might have been descendants of progymnosperms.

The rise and early differentiation of gymnosperms were clearly associated with EB, where in the uppermost Devonian - Lower Carboniferous the findings of seeds are fairly common. Also in EB, the gymnosperms very early diverged into the classes Ginkgopsida and Cycadopsida (sensu Meyen /8/). The distribution of the orders Calamopityales and Callistophytales (class Ginkgosida) is difficult to trace among phytochoria. The Calamopityales could have inhabited, beside EP, also Gondwanaland, and in this area they seem to have given rise to the order Arberiales (glossopterids) which are so characteristic for the Gondwana Upper Palaeozoic. Callistophytales are known outside EB in the Middle-Upper Carboniferous of Angaraland (Paragondwanidium-Gondwanotheca-Angarocarpus).

The transition from the Callistophytales to Peltaspermales occured in EB or in its frontier regions with Western Angaraland. In the Early Permian the Peltaspermales were widespread in west of Eurasia and in North America. In the Late Permian their families Peltaspermaceae and Cardiolepidaceae prevailed in the Subangara area (western and southern periphery of Angara kingdom covering previously recognized East-European and Ural-Kazakhstan areas along with adjacent regions /7/), and the Peltaspermaceae are also known in the Atlantic kingdom. The gigantopterids characteristic of the Permian of Cathaysia and North America probably belong to the Peltaspermales and, partly, Callistophytales. At the end of the Permian - beginning of the Triassic, the Peltaspermales penetrated into Gondwanaland, wherein they soon became the dominating plant group (family Umkomasiaceae = Corystospermaceae). In the Jurassic the Peltaspermales (Rhaphidopteris, Pachypteris) are characteristic of EB (chiefly for west of Eurasia), but rarely occur in Gondwanaland. In the Triassic the Peltaspermales gave rise to the Caytoniales, which are mostly confined to EB and at times penetrated into the Siberian area. The Peltaspermales were also ancestors of Ginkgoales and Leptostrobales /8, 9/. The transition of these orders happened in the northern periphery of EB or in the Subangara area; thereafter the ginkgos became an ubiquitous group. They are less common in EB and rare in Gondwana, but prevail in the Siberian area in the Jurassic and Early Cretaceous, together with Leptostrobales (czekanowskians). The latter are common in EB only in places, and totally absent in Gondwanaland.

The order Cordaitanthales (cordaites) became separated from the Lagenostomales in the Early Carboniferous. As early as the beginning of the Middle Carboniferous the order was divided into three lineages. One of them (Cordaitanthaceae) evolved within EB (or in ecotone zone with Angaraland) where it gave rise to another order of Pinopsida, viz. the conifers (order Pinales s.l.). Two other lineages (Rufloriaceae, Vojnovskyaceae) are unknown outside Angaraland (their records in Gondwana and North America are erroneous). Both the Angara families preserved archaic characters (leaf-like habit of scales in the axillary complex; profusely branched microsporoclads in Kuznetskia; etc.), but some characters are fairly advanced (e.g. coniferalean-like strobili of Cladostrobus). The presence of Cordaitanthales in Gondwana has not been supported with the plants at a given taxonomic level (from family upwards) is confined to the EB. The humid portions of the EB always exhibit the maximum diversity in floras, but this does not mean that these very areas were centres of formations of all higher taxa. In some cases new taxa of higher rank might have evolved in arid or semiarid (non-coal bearing) areas of the EB and adjacent ecotones. For instance, the Subangara area of the Permian might have been an important florogenetic territory.

In extraequatorial areas only two endemic orders are known (Arberiales and Pentoxylales in Gondwana; although the Pentoxylales appeared in Gondwana when its climate could actually have been more or less equatorial), and a few endemic families. In general, the florogenetic contribution of phytochorias decreases towards the poles. Obviously, the abiotic factors of selection in extraequatorial areas cut off any significant evolutionary innovations. It is suggestive, in this respect, that many morphological features occuring in modern temperate plants as abnormalities rapidly eliminated by selection, are preserved in the tropics where they occur as a norm in closely related taxa /14/.

5. The above mentioned phenomenon of extraequatorial persistence can provoke errors in both stratigraphic correlations and palaeoclimatic reconstructions. The presence of residual, conservative migrants gives a false appearance of greater age to the assemblages containing them. If the persistent forms of an equatorial origin have survived until the present day in extraequatorial floras, their presence in older floras of the EB may lead to the reconstruction of a cooler climate then it was in reality.

6. In palynological assemblages the earliest representatives of higher taxa are often recorded one stage earlier than in plant megafossil assemblages, even when the collecting was very extensive. This observation may constitute evidence that major

a fairly rapid dedifferentiation of the terrestrial vegetation occured. Then a new phase of differentiation emerges, continuing until the present time. The main homogenizing factor in the distribution of taxa on the Earth surface now is mankind.

2. Plant assemblages within each phytochoria evolve in accordance with a model of punctuated equilibrium. The periods of stasis are replaced by rapid changes involving the composition of assemblages or the proportion between their components. The rearrangements in assemblages in extraequatorial areas, especially on a regional scale, appear to be mostly connected with temperature fluctuations, and, less frequently, fluctuations between aridity and humidity. In the EB the changes in temperature regime seem to play a less important role. Periods of stasis serve to establishing regional and interregional phytostratigraphical units, whereas climatic events may serve as a basis for interregional and global stratigraphic correlation.

3. Climatic events often involve only rearrangements of local assemblages and do not disturb their overall autochtonous development. In other cases climatic changes provoke migrations of plants from adjacent phytochorias. During cooling of climate, extraequatorial plants do not penetrate into the EB as solitary migrants, but rather all the extraequatorial biota moves as a whole towards the Equator. In contrast, warmings of climate allow some equatorial plants to penetrate into extraequatorial areas. These migrants either vanish during the next cooling, or persist over a long period, preserving their archaic habit and undergoing but slight evolutionary modifications (the phenomenon of extraequatorial persistence. Only rarely do these migrants give rise to new higher taxa.

4. If suprageneric and generic taxa are cosmopolitan or nearly so, they appear first in the EB and only later in extraequatorial areas. The reverse situation, in taxa ranging from family upwards, is unknown. One can conclude that the main diversification of higher

Cycadales and Bennettitales might have been derived. The oldest representatives of these two orders are known in the Triassic and are confined to the EB. Both the orders could have come from Cathaysia, where cycadophyte leaves are fairly common in the Permian. In the Jurassic - Cretaceous the cycadophytes are particularly numerous in the EB and Gondwanaland, whereas in the Siberian area they mostly occur in intervals corresponding to warmings of climate. Each genus recorded in Siberia appears earlier in the EB.

Comparative evolutionary analysis of gymnosperm and angiosperm reproductive organs gives convincing evidence /8/ that the ancestors of angiosperms belonged to the class Cycadopsida (the order is unknown). Judging from palynological observations /2/ angiosperms appeared at first (in the Barremian) and in the largest quantities in regions adjoining the present-day Atlantic oceans. The regions correspond to low palaeolatitudes and belong to the arid or semiarid palaeoclimatic zone, i.e. they belong to the EB. Quite rapidly (in the Aptian-Albian) angiosperms became cosmopolitan. Older hypotheses on the Arctic origin of angiosperms can be rejected.

Such is, in very broad features, the stratigraphic and geographic distribution of major phylogenetic lineages of higher plants. The above analysis along with observations on plant assemblages in phytochorias allows us to offer the following tentative conclusions (working hypotheses).

1. In the evolution of terrestrial vegetation two major cycles of differentiation and dedifferentiation were long established /12, 13/. A distinct pattern of phytochoria established on higher plants appeared at the beginning of the Early Carboniferous, the differentiation of the floras being developed faster than that of marine fauna. The maximum differences between phytochoria, and the maximum number of phytochoria of different rank were attained in the Late Permian. At the very end of the Permian - beginning the Triassic,

discovery of fructifications. One cannot exclude the possibility that the Gondwana Cordaites leaves belong to the Arberiales.

The Palaeozoic evolution of conifers proceeded mainly in more arid regions of EB and later in the Subangara area. In the Euramerian area as early as the beginning of the Permian the axillary complex was transformed into a primitive seed scale with vestigial accompanying sterile scales. The later forms of the same lineage (Voltziaceae) diverged into the families Araucariaceae, Podocarpaceae, Taxodiaceae, Pinaceae, Cheirolepidiaceae and Cycadocarpidiaceae during the Triassic or at the very beginning of the Jurassic. The evolution of Araucariaceae further proceeded in EP and Gondwana. Pinaceae and Taxodiaceae were widely spread outside Gondwana. Podocarpaceae were nearly ubiquitous in the Jurassic -Cretaceous, but later their evolution was restricted to Gondwana. Cheirolepidiaceae survived until the Eocene and were particularly numerous in aridsemiarid equatorial areas penetrating towards both poles in intervals of warming and aridization. Typical Cycadocarpidiaceae (showing fructifications) are recorded only in EB, but the leaf genus Podozamites affiliated with the family was ubiquitous in the Jurassic and Cretaceous.

The class Cycadopsida (sensu Meyen /8/) was initially represented by a single order Lagenostomales reliably recorded only in EB, where, approximately in the Visean, it gave rise to the order Trigonocarpales. The main divergence of the latter proceeded also in EB. The Trigonocarpales may cover the plexus of Angara plants like Angaropteridium-Abacanidium-Cardiopteridium appearing in the Visean. Trigonocarpaleans with foliage of the Neuropteris type persisted in Cathaysia and Angaraland till the Late Permian. The Angara forms produced paripinnate fronds. The paripinnate neuropterids became extinct in Euramerian area much earlier (before the end of the Middle Carboniferous). The evolutionary fate of the Trigonocarpales is unknown, but this is the only order of Cycadopsida from which the orders evolutionary innovations did not happen directly in sedimentary basins (e.g. coal producing swamps), but at some distance, probably on adjacent areas of greater elevation.

7. There is no evident dependency between the intensity of phylogenetic diversification and a certain degree of floristic differentiation. The appearance of new taxa of the rank of family and upwards are evenly distributed throughout the geochoronological scale. The intensity of diversification passed in due course from division or class of higher plants to others. The very absence of a periodicity in the formation of higher taxa allows us to conclude that the phylogeny of higher plants as itself cannot serve for the subdivision of the geological history and, hence, for the substantiation of the standard geochronological units and their boundaries. Only florogenesis as a whole can serve this purpose.

The author is grateful to Prof. W.G. Chaloner (London) for going through the manuscript and correcting the English.

LITERATURE

1. Chaloner, W.G. & Meyen, S.V. 1973. Carboniferous and Permian floras of the Northern continents. In: A. Hallam (ed.) Atlas of palaeobiogeography. Elsevier, Amsterdam et al., p. 169-186.
2. Doyle, J.A., Biens, P., Doerenkamp, A. & Jardine, S. 1977. Angiosperm pollen flora from the pre-Albian Lower Cretaceous of Equatorial Africa. Bull. Cent. Rech. Explor.-Prod. Elf-Aquitaine, t. 1, No. 2. p. 541-473.
3. Edwards, D. 1973. Devonian floras. In: A. Hallam (ed.) Atlas of palaeobiogeography. Elsevier, Amsterdam et al., p. 105-115.
4. Gothan, W. & Weyland, H. 1973. Lehrbuch der Palaobotanik. Akad. Verlag, Berlin, 677 S.

5. Kryshtofovich, A.N. 1957. Palaeobotany. Gostoptekhizdat, Leningrad, 650 p. (in Russian).
6. Kryshtofovich, A.N. 1959. Selected works. Vol. I. Theoretical works. Izdat. Akad. Nauk SSSR, Moscow-Lenningrad, 510 p. (in Russian).
7. Meyen, S.V. 1981. Vestiges of Indian herbs. Mysl, Moscow, 159 p. (in Russian).
8. Meyen, S.V. 1982. Gymnosperm fructifications and their evolution as evidence by palaeobotany. Zhurn. Obshch. Biol., vol. 43, No. 3, p. 303-323 (in Russian, with English summary).
9. Meyen, S.V. 1983. The systematics of peltaspermaceous pteridosperms and their position in the phylogeny of gymnosperms. Bull. Moskovsk. Obshch. ispyt. prirody, otd. biol., vol. 88, No. 1, p. 3-14 (in Russian, with English summary).
10. Schopf, J.M. 1973. The contrasting plant assemblages from the Permian and Triassic deposits on southern continents. In: A. Logan & L. Hills (eds.). The Permian and Triassic systems and their mutual boundary. Alberta Soc. Pet. Geol., Mem. 2, p. 379-397.
11. Seward, A.C. 1933. Plant life through the ages. Cambridge Univ. Press, 601 p. (Russian edition of 1936).
12. Vakhrameev, V.A., Dobruskina, I.A., Zaklinskaya, E.D., Meyen, S.V. 1970. Palaeozoic and Mesozoic floras of Eurasia and phytogeography of this time. Trudy Geol. Inst. Akad. Nauk SSSR, vyp. 208, 426 p. (in Russian).
13. Vakhrameev, V.A., Dobruskina, I.A., Meyen, S.V., Zaklinskaja, E.D. 1978. Palaozoische und mesozoische Floren Eurasiens und die Phytogeographie dieser Zeit. Gustav Fischer Verl., Jena, 300 S.
14. Van Steenis, C.G.G.J. 1969. Plant speciation in Malasia with special reference to the theory of non-adaptive, saltatory evolution. Bot. J. Linn. Soc., vol. 1, p. 97-133.
15. Zimmermann, W. 1959. Die phylogenie der Pflanzen. Gustav Fischer Verl., Stuttgart, 777 S.

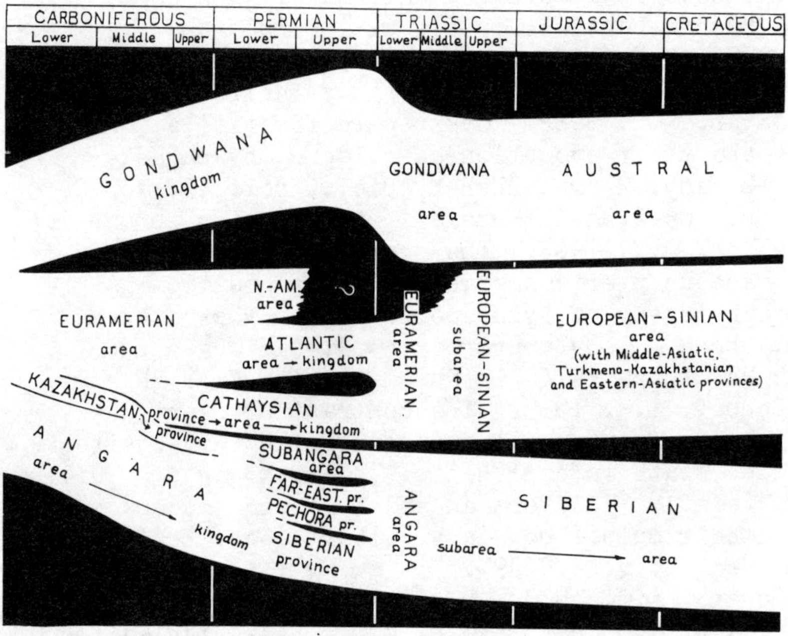

FIGURE 1. Evolution of phytochoria in Late Pálaeozoic and Mesozoic. N.-AM. - North-American. This area together with Atlantic and Cathaysian areas constitute Amerosinian Kingdom of the Early Permian. The Kingdom was preceded in the Carboniferous by Arctocarbonic kingdom consisting of Euramerian and Angara areas. Far-Eastern, Pechora and Siberian provinces belong to Angara area which, together with Subangara area constitute Angara kingdom.

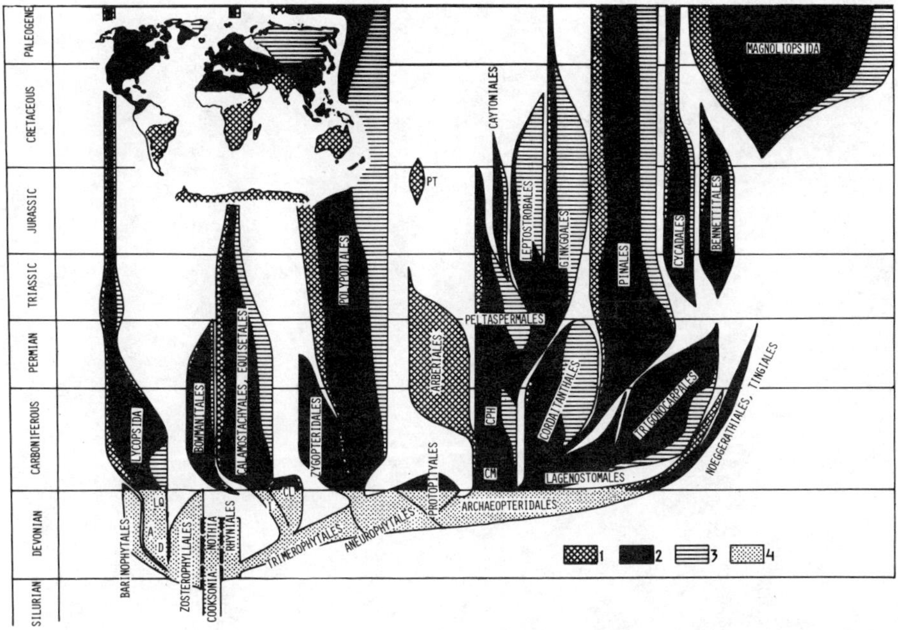

FIGURE 2. Phylogeny of vascular plants and geographic pattern of main phyla. Map on inset shows basic floristic zonation in Late Palaeozoic (time of principal divergency of gymnosperm phyla). D - Drepanophycus; A - Asteroxylon; LQ - Leclercqia; CL - Cladoxylales; I - Ibykales (= Iridopteridales); CM - Calamopityales; CPH - Callistophytales; PT - Pentoxylales; 1 - Nothal (Gondwana) extraequatorial areas; 2 - equatorial and adjacent ecotonal areas; 3 - boreal extraequatorial areas; 4 - uncertain phytogeographic situation. Width of phylogenetic branches mostly reflects role of corresponding taxa in plant assemblages of pertinent time span.

ULTRASTRUCTURAL RESEARCH ON MOLLUSCAN NACRE

HARRY MUTVEI

Swedish Museum of Natural History, S-104 05 Stockholm, Sweden.

ABSTRACT

There is a considerable ultrastructural similarity between the nacreous layer and the spherulitic-prismatic layer in molluscan shells. Both are composed of acicular crystalline elements which in the nacre are oriented strictly vertically whereas they have a varying orientation in the spherulites and prisms. In both layers there are numerous, parallel, regularly spaced, organic membranes which in the nacre are thicker and form the interlamellar membranes. In bivalves, gastropods and *Nautilus* the crystalline elements are arranged to form polygonal tablets. This nacre is the nacreous Type 1. In the dibranchiate cephalopods *Spirula*, *Sepia* and some fossil belemnoids the crystalline elements form parallel rods. This is the nacreous Type 2. In the nacreous Type 1 the tablets show complex twinning which is more regular in bivalves than in gastropods and *Nautilus*. In bivalves the twinning pattern in the preceding tablet determinates the nucleation site of a new tablet. In gastropods and *Nautilus*, on the other hand, a conspicuous accumulation of calcified organic matrix in the centre of the preceding tablet determinates the nucleation site of a new tablet. There is a crystallographic continuity between the consecutive mineral lamellae of the Type 1 through the interlamellar organic membranes. The function of these membranes is still obscure.

Proceedings of the 27th International
Geological Congress, Volume 2, pp. 111–124
PALAEONTOLOGY

INTRODUCTION

An understanding of the mechanism of biomineralization is essential for studies of (1) hard tissues in plants, invertebrates and vertebrates, and (2) origin and lithification of biogenic sediments. A review on the multitude of aspects on biomineralization and biological heavy metal accumulation has been recently published in the papers presented at the Fourth International Symposium on Biomineralization (1983, eds: Westbroek and de Jong).

The nacreous layer in molluscan shells has been extensively studied for the understanding of the mechanism of biomineralization. Despite these studies, this mechanism is still imperfectly understood. Below, some of the aspects on the current research will be reviewed and discussed.

MINERAL COMPONENTS

Two structurally different types, Type 1 and 2, have been distinguished in molluscan nacre (Mutvei, 1970). The nacre in the shells of recent monoplacophors, gastropods, bivalves, and the cephalopod *Nautilus* belongs to Type 1. In this type, the mineral components form polygonal aragonite tablets. The morphology and arrangement of the tablets have been studied by numerous authors, e.g. Grégoire, 1957, 1962, 1967; Wada, 1961, 1972; Wise, 1970; Erben,1972; Towe and Hamilton, 1968.

Early studies gave the result that the aragonite tablets are single crystal units. By using various methods of etching and demineralization it became apparent that the tablets, on the contrary, have a highly complex mineral

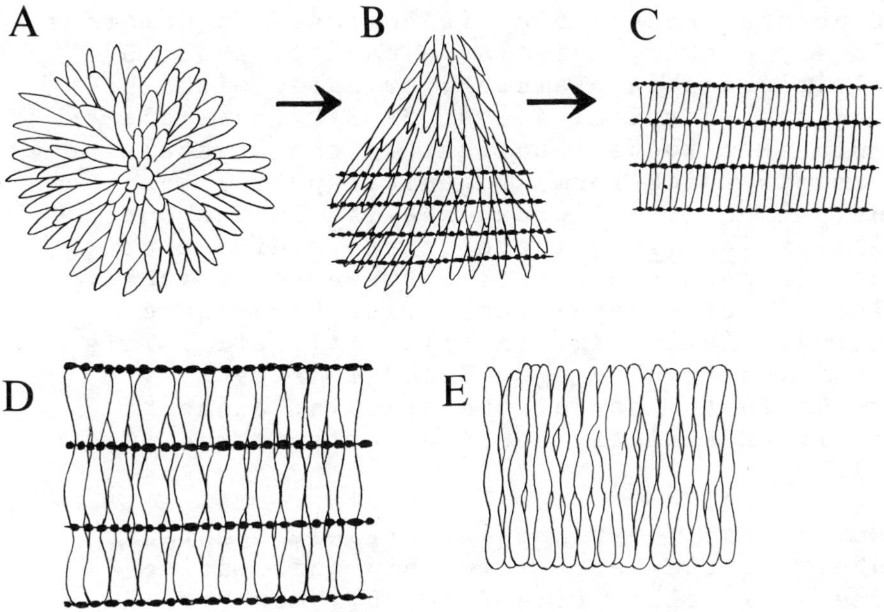

Figure 1. A, aragonite spherulite with spherically oriented acicular crystalline elements. B, aragonite prism formed as a sector of a spherulite. Lower portion of the prism has a semi-nacreous structure due to the occurrence of typical interlamellar organic membranes. C, vertical section of typical nacre with interlamellar organic membranes and vertically oriented crystalline elements in the mineral lamellae. D, vertical section of nacre treated with a glutaraldehyde solution. E, the same section treated subsequently with a sodium hypochlorite solution. Note the continuity of the acicular crystalline elements between the consecutive mineral lamellae.

structure. Each tablet is composed of numerous, closely packed, acicular elements, about 0.1 millimicrons in diameter, oriented with their long axes vertical (figs. 2A B). These elements seem to be identical to the crystalline elements which form aragonite spherulites and prisms in molluscan shells. In the shell wall of Nautilus, there exists a direct crystallographic continuity between the acicular elements of spherulites, prisms and nacre (Mutvei, 1972, fig. 1A, pls. 1:1, 2:1). This condition is here elucidated in Figs. 1A, B and C. In the nacre, the acicular elements acquire a strictly vertical orientation (fig. 1C).

Horizontal interlamellar organic membranes subdivide the nacre into thin consecutive mineral lamellae (figs. 1C, D). However, numerous acicular crystalline elements continue through the interlamellar membranes, and thus maintain a crystallographic continuity between the nacreous tablets in consecutive mineral lamellae (compare figs. 1D and 1E; see also Mutvei, 1979, figs. 6, 7, 8).

The formation of the polygonal aragonite tablets, characteristic for the nacreous type 1, is related to the twinning of the aragonite. The following two twinning patterns were found to characterize the nacre in different molluscan classes:

(1) In the bivalves hitherto studied (Mutvei, 1977, 1979, 1980), the tablets have the most complex twinning pattern amongst the molluscs. Each tablet is composed of two layers: a thick basal layer, and a thin upper layer (fig. 2A). In the basal layer, the acicular crystalline elements have an uniform arrangement in that they form numerous parallel rows (fig. 2A).

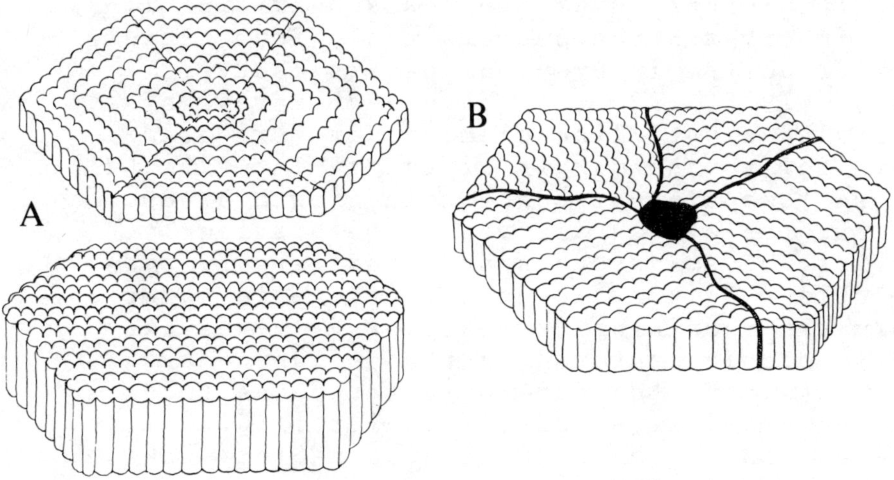

Figure 2. A, structure of a nacreous tablet in a recent bivalve. Note the two crystalline layers. In the basal layer, the acicular crystalline elements form parallel rows. The upper layer is subdivided into six crystalline sectors. In each sector, the rows of acicular elements are parallel to the margin of the tablet. B, structure of nacreous tablet in recent gastropods and *Nautilus*. Note the subdivision of the tablet into crystalline sectors separated by vertical organic membranes. A conspicuous accumulation of calcified organic matrix (black) occurs in the centre of the tablet.

The thin upper layer shows a subdivision into six crystalline sectors (only four sectors were reported in previous papers, Mutvei, 1977, 1978, 1980). In each sector rows of acicular crystalline elements are oriented parallel to the margin of the tablet (fig. 2A). These rows are not concentric growth bands of the tablet, as assumed by me previously. Two of the six sectors are considerably less soluble in glutaraldehyde-acetic acid solution than the remaining four sectors. In the two less-soluble sectors the rows of acicular elements have the same orientation as those in the basal layer of the tablet (fig. 2A). The nucleation of new nacreous tablets takes place invariably on the surfaces of the less-soluble sectors (Mutvei, 1977, figs. 7, 9, 10A-C).

(2) In the gastropods hitherto studied, and in Nautilus, the twinning pattern of the nacreous tablets is less regular, and in many ways similar to that in inorganically formed aragonite "crystals" (Mutvei, 1978, 1980). Only a single mineral layer can be distinguished in each tablet. This layer is subdivided into a varying number of crystalline sectors which form interpenetrating and cyclic twins (fig. 2B). The adjacent sectors are separated from each other by vertical organic membranes (fig. 2B). In each sector, the acicular crystalline elements form parallel rows with different orientation in different sectors (fig. 2B). The most prominent and constant feature is the accumulation of a calcified organic matrix in the centre of each tablet (fig. 2B; see also Mutvei, 1978, 1980). On this matrix, the nucleation of new nacreous tablets takes place, and as a result, the consecutive nacreous tablets acquire a columnar arrangement (columnar nacre).

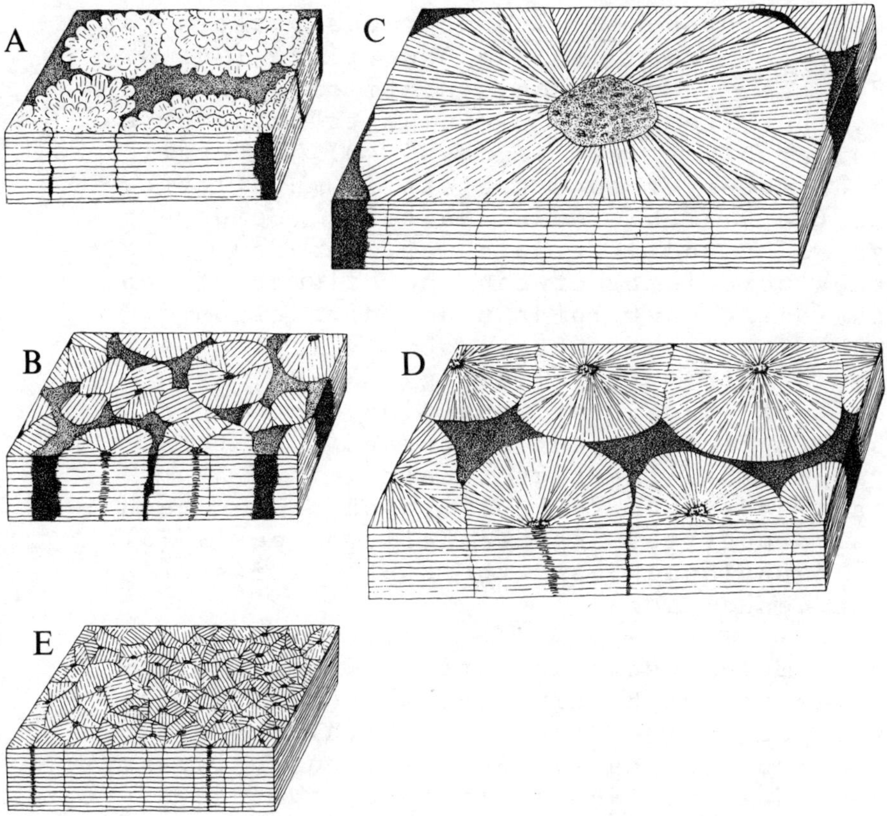

Figure 3. A, B, C, nacreous layer of the shell wall in a Lower Ordovician endoceratid cephalopod from the Island of Öland, Sweden. Note the variations in the size and shape of the nacreous tablets. The interspaces between the columns of tablets were probably filled by an organic matrix (black). D, Isorthoceras from the Upper Ordovician Maquoketa shale, USA. Note the large, regularly growing nacreous tablets and the interspaces between the columns of tablets. E, recent Nautilus. The nacre of the shell wall is composed of vertical columns of small sized tablets, without interspaces.

The gastropod-*Nautilus* nacreous pattern was developed already in the Ordovician (Mutvei, 1983a, b). In orthoconic nautiloids of this age, the size of the nacreous tablets varies considerably (figs. 3A, B, C, D). However, the diameter of most tablets is much larger than that in the recent *Nautilus* (compare figs. 3A-D and 3E). Moreover, numerous interspaces were left between the columns of tablets in the nautiloids under discussion. These interspaces were probably filled by an organic matrix (dark in figs. 3A-D). As a result of the latter condition, the nacre was more flexible but possessed less tensile strength than the recent nacre. This has a great importance for estimation of the mechanical properties in the shells of early Paleozoic cephalopods, and the depth limits of these cephalopods.

The complex nacreous pattern which characterizes recent bivalves seems not to have developed in the Ordovician bivalves from the Maquoketa shale, USA, previously investigated (Mutvei, 1983a). Instead, the nacre shows an irregular pattern.

The nacreous Type 2 (Mutvei, 1970; "lamello--fibrous", Erben, 1972) differs from the Type 1 by the fact that the nacreous lamellae are not composed of polygonal tablets but of thin parallel crystalline rods. This nacreous type occurs in the septa of shell secreting dibranchiata cephalopods: Recent *Spirula* and *Sepia*, and fossil belemnoids (Dauphin and Keller, 1982). It occurs also in the dorsal shield of *Sepia* and in the wall of the phragmocone in certain belemnoids, whereas in other belemnoids the nacreous Type 1 may occur in this wall (Barskov, 1981, personal communication; Dauphin and Keller, 1982;

Dauphin, 1983; Mutvei, 1982, unpublished observations).

It is still difficult to compare in detail the nacreous Types 1 and 2. During the evolution, the nacreous Type 1 seems to become structurally modified in the septa of nautiloid and ammonoid cephalopods (unpublished observations) Type 2 may correspond to the rows of acicular crystalline elements in the nacreous Type 1.

ORGANIC COMPONENTS

The most conspicuous organic components in the nacreous Type 1 are the interlamellar organic membranes which separate the consecutive mineral lamellae. According to Erben and Watabe (1974) and Nakahara (1981) for example, these membranes originate as a result of a fusion between (1) horizontal organic sheets formed before the precipitation of $CaCO_3$, and (2) organic envelopes around the growing nacreous tablets (see also Grégoire, 1957). The adjacent tablets in each lamella are separated by vertical membranes which are parts of the envelopes. However, it must be stressed that the occurrence of the regularily spaced, parallel organic membranes is by no means exclusive for the nacreous layer. Such membranes occur also frequently, if not always, in the prismatic and spherulitic shell layers but they are here considerably thinner (unpublished observations by Y. Dauphin and H. Mutvei).

The structure of the interlamellar organic membranes and its taxonomic significance have been studied by several authors, e.g. Grégoire (1957, 1960, 1961, 1962, 1967, 1982), Mutvei (1969), Iwata (1975), Goffinet _et al_.

(1977). Each membrane usually consists of thick trabeculae and thin intertrabecular areas, both of which are insoluble in EDTA.

An EDTA soluble organic matrix has been reported to occur within the nacreous tablets. This was termed the "intracrystalline" fraction, and it was supposed to have an irregular spatial distribution within the tablets. However, as recently demonstrated (Mutvei, 1977, 1978, 1980), the latter fraction (intra-tabletal fraction)forms membranes around the crystalline elements, and consequently it has a highly regular distribution.

In the nacreous layer of recent bivalves two sets of membranes can be distinguished in the intratabletal fraction: (1) membranes which encase each of the vertical acicular mineral elements (Mutvei, 1979; fig. 9), and (2) horizontal membranes which separate the thick basal layer of the tablet from the thin upper layer (Mutvei, 1979; fig. 4, 5). In the gastropod-Nautilus nacre at least three intratabletal organic components can be distinguished: (1) membranes which encase each of the vertical acicular mineral elements; (2) vertical radial membranes which separate the adjacent crystalline sectors of the tablet; and (3) a conspicuous accumulation of calcified organic matrix housed in the central portion of the tablet (fig. 2B; see also Mutvei, 1978). The fraction (1) seems to have a different solubility than the fractions (2) and (3).

In the nacreous Type 2 only one type of organic membranes has been previously recognized. These membranes encase the mineral rods and form the horizontal interlamellar membranes (Grégoire, 1961; Mutvei, 1969, 1970).

It has been suggested by several authors that either the soluble or the insoluble organic matrix may serve as a template for calcium carbonate nucleation. This is probably the case in the gastropod-*Nautilus* nacre where the conspicuous accumulation of a calcified organic matrix in the centre of the tablet determinates the nucleation site of a new tablet. The interlamellar organic membrane is structurally modified at this place forming the so-called transparent elevation (Mutvei, 1969; Goffinet *et al*., 1977). A quite different nucleation mechanism is found in the recent bivalve nacre. Here, the nucleation of a new tablet takes place invariably on the surface of the less soluble pair of twin crystals in the preceding tablet. These crystals are characterized only by the fact that the orientation of rows of acicular crystalline elements coincides with that in the principal basal layer of the tablet. Consequently, the twinning pattern in the preceding tablet determinates the nucleation site of a new tablet in bivalves. Thus, the nucleation of new tablets seems not to be related to the chemical and structure properties of the interlamellar membranes. The function of these membranes still remains obscure.

REFERENCES

Dauphin, Y. (1983). Microstructure du genre triasique *Aulacoceras* (Cephalopoda-Coleoidea). Remarques sur les homologies des tissus coquilliers chez les Céphalopodes. N. Jb. Geol. Palaeont. Abh., 165, 418-437.

Dauphin, Y. and Keller, J.P. (1982). Mise en évidence d'un type microstructural coquillier spécifique des Céphalopodes dibranchiaux.

C.R. Acad. Paris 294, 404-412.

Erben, H.K. (1972). Uber die Bildung und das Wachstum von Perlmutt. Biomineralization 4, 15-46.

Erben, H.K. and Watabe, N. (1974). Crystal formation and growth in bivalve nacre. Nature (London) 248, 128-130.

Goffinet, G., Grégoire, Ch. and Voss-Foucart, M.F. (1977). On ultrastructure of the trabeculae in the interlamellar membranes of nacre conchiolin of the *Nautilus* shell. Archiv. Intern. Physiol. Biochem. 85, 849-863.

Grégoire, Ch. (1957). Topography of the organic components in the mother-of-pearl. Jour. Biophys. Biochem. Cytology, 797-808.

Grégoire, Ch. (1960). Further studies on structure of the organic components in mother-of-pearl, especially in pelecypods. Bull. Inst. Roy. Sci. Natur. Belg. 36, 1-22.

Grégoire, Ch. (1961). Sur la structure submicroscopique de mollusques. Bull. Inst. Roy. Sci. Natur. Belg. 37, 1-37.

Grégoire, Ch. (1967). Sur la structure des matrices organiques des coquilles de mollusques. Biol. Rev. 42, 653-688.

Grégoire, Ch. (1972). Ultrastructure des composants organiques des coquilles de mollusques. Haliotis 2, 51-79.

Iwata, K. (1975). Ultrastructure of the conchiolin matrices in molluscan nacreous layer. Jour. Fac. Sci. Hokkaido Univ. 17, 173-229.

Mutvei, H. (1969). On the micro- and ultra-structure of the conchiolin in the nacreous layer of some recent and fossil molluscs. Stockholm Contr. Geol. 20, 1-17.

Mutvei, H. (1970). Ultrastructure of the mineral and organic components of molluscan nacreous layers. Biomineralization 2, 49-72.

Mutvei, H. (1972). Ultrastructural relationships between the prismatic and nacreous layers in _Nautilus_ (Cephalopoda). Biomineralization 4, 81-86.

Mutvei, H. (1977). The nacreous layer in _Mytilus_, _Nucula_ and _Unio_ (Bivalvia). Calcif. Tiss. Res. 24, 11-18.

Mutvei, H. (1978). Ultrastructural characteristics of the nacre in some gastropods. Zoologica Scripta 7, 287-296.

Mutvei, H. (1979). On the internal structures of the nacreous tablets in molluscan shells. Scanning Electron Microscopy 1979/2, 457-462.

Mutvei, H. (1980). The nacreous layer in molluscan shells. In: Proc. 3rd Intern. Biomin. Symp., M. Omori and N. Watabe (eds). Tokai Univ. Press, Tokyo, pp. 49-56.

Mutvei, H. (1983a). Ultrastructural evolution of molluscan nacre. In: Biomineralization and biological metal accumulation, P. Westbroek and E.W. de Jong (eds). Reidel Publ. Comp., Dordrecht, pp. 267-271.

Mutvei, H. (1983b). Flexible nacre in _Isorthoceras_ (Cephalopoda: Nautiloidea), with remarks on the evolution of cephalopod nacre. Lethaia 16, 233-240.

Nakahara, H. (1981). The formation and fine structure of the organic phase of the nacreous layer in mollusc shell. Study of Molluscan Paleobiology (Prof. Omori Mem. Vol.), Niigita Univ., Japan, pp. 21-27.

Towe, K.M. and Hamilton, G.H. (1968). Ultrastructure and inferred calcification of the mature and developing nacre in bivalve mollusks. Calcif. Tiss. Res. 1, 306-318.

Wada, K. (1961). Crystal growth of molluscan shells. Bull. Natl. Pearl Res. Lab. 7, 703-828.

Wada, K. (1972). Nucleation and growth of aragonite crystals in the nacre of some bivalve molluscs. Biomineralization 6, 141-159.

Wise, S.W., (1970). Microarchitecture and mode of formation of nacre (mother-of-pearl) in pelecypods, gastropods, and cephalopods. Ecologae geol. Helv. 63, 775-797.

CAMBRIAN EVOLUTIONARY STUDIES: CONSTRAINTS AND CHALLENGES

ALLISON R. PALMER
Geological Society of America, Box 9140, Boulder, CO.,USA

ABSTRACT

Effective studies of evolution, migration and biogeography of Cambrian faunas are dependent on geologically defensible paleogeography, a reliable body of information about paleooceanography, detailed biostratigraphy and biofacies analysis, careful modern systematics and an understanding of the causes of widespread intracambrian extinction events. Information on all of these topics is incomplete or lacking on a global scale. Thus, most scenarios about evolution, migration and biogeography of Cambrian faunas on this scale are likely to have very little substance. Reliability of interpretations in these topic areas is indirectly proportional to the size of the region that is considered.

INTRODUCTION

Somewhat over half a billion years ago, the metazoa of the world's oceans began to develop mineralized exoskeletons, first predominantly phosphatic, and subsequently predominantly calcareous. The record of the first 65 to 75 million years of development of this marine biota is preserved in the Cambrian system.

The body of data that is potentially available for study of the evolution, migration and geographic fractionation of the largely extinct classes of Cambrian invertebrates comes from a fragmentary record preserved in a geographic framework that bears little resemblance to the geographic framework in existence when the organisms were alive. Thus, a major prerequisite for effective large scale evolutionary studies of Cambrian organisms is a proper context.

We need first to determine a reliable Cambrian geography; we also need to develop a body of information about

Proceedings of the 27th International
Geological Congress, Volume 2, pp. 125–134
PALAEONTOLOGY

major aspects of Cambrian paleooceanography and climate--ocean currents, wind systems, sea level fluctuations, and properties of water masses such as temperature and salinity; and finally we need to find an explanation for the cause or causes of several well-documented, abrupt, and widespread extinction events that perturbed the regular progression of evolutionary change in Cambrian time. To the degree that our information about these topics is inadequate, we are inhibited in our ability to speculate effectively about major evolutionary questions involving the Cambrian biota,

THE PALEOGEOGRAPHIC CHALLENGE

We do not yet have an unequivocal reconstruction of Cambrian paleogeography. The most recent attempts (Scotese and others, 1979, p. 221; figs. 2-7; Ziegler and others, 1979, p. 483-485) may be reliable for the four undisputed continents discussed below, but identification of the original locations of the many suspect or exotic terranes that contain important occurrences of Cambrian rocks and fossils has so few present constraints that it remains as an opportunity for further research and a frustration for studies of biogeography and faunal migration.

General agreement seems to have been reached about four major Cambrian Continents: Laurentia--North America minus its eastern seaboard and very probably most of what is now Mexico and Central America; Baltica--from the Caledonides to the Urals, and bounded on the south by the Tornquist line and the Hercynian orogen; Siberia--from the west-Siberian lowlands to at least the Kolyma Platform, and from the Taimyr foldbelt to the south Siberian foldbelt; and Gondwana--including most of the modern continents of Africa, South America, Antarctica, and Australia, and the Indian subcontinent.

Regional lithofacies patterns on each of these Cambrian continents in most instances confirm their surrounding oceanic regions and central cratons, although there are key gaps in information for Gondwana along its northern South American and North African margins and between Africa and Australia.

A few active and passive continental margins of Cambrian age can be identified. These reveal information about plate tectonic activities during Cambrian time. Active tectonic margins of Cambrian age are clearly indicated for southern Siberia, eastern Australia and its continuation into Antarctica, and western Baltica. Passive margins of Cambrian age are well defined around most of Laruentia and eastern and northern Siberia. The tectonic state of most other Cambrian continental margins is not completely clear.

A number of areas with significant Cambrian outcrops are isolated from the major Cambrian continents by younger Phanerozoic orogenic belts. These appear from their preserved stratigraphy to be fragments of Cambrian lands that were broken up and disseminated in post-Cambrian time. The later Phanerozoic history of these terranes may provide the clues necessary to determine the times of their breakup and subsequent docking. Their internal litho- and biostratigraphy will provide important clues about their origin.

Cambrian terranes that are at least suspect, and most likely exotic, are found in parts of the present North Atlantic borderlands, southern Mexico, northern Alaska, and most of southern Eurasia from France and Spain to China and Malaysia--including much of the southern part of Asiatic USSR and almost all of the Middle East.

Some assistance with the exotic or suspect terrane problem is available. In Kazakhstan and Yunnan, high grade phosphorites are associated with the Precambrian/Cambrian boundary, and overlying Early Cambrian faunas include redlichiid and eodiscid trilobites of Chinese aspect (Ergaliev and Pokrovskaya, 1977). The later Middle and Upper Cambrian faunas in southern Kazakhstan (Ergaliev, 1980) include many genera shared with China and Australia. Further ties from southern Asia to the Australian parts of Gondwana include the occurrence of *Xystridura* in the Middle Cambrian of Hainan Island (Sun, 1963) associated with phosphorites, analogous to the medial Middle Cambrian *Xystridura*-phosphorite associations of Queensland; and the inner detrital belt red sandstones of western Malaysia with late Cambrian suakiid trilobites that can be closely matched to similar lithologies and faunas in the Northern Territory of Australia. Both faunal and lithologic information suggest close ties between the suspect or exotic terranes of Kazakhstan, southeastern Asia and the Australian sector of Gondwana.

The thick Late Proterozoic/Early Cambrian salt deposits of Iran and Pakistan can be correlated with the Arabian part of Gondwana (Wolfart, 1983, p. 25) and scattered Cambrian faunas of the Middle East have strong Chinese/ Australian affinities. Although all of these observations suggest that many southern Eurasian terranes should be relocated along the African-Indian-Australian sector of Gondwana, much work remains to be done on this and alternative hypotheses (e.g. "Kazakhstania", Ziegler and others, 1979, p. 484; Scotese and others, 1979, figs. 2-7).

In the exotic terranes of eastern Canada and northeastern United States, the faunas and stratigraphy of the Early and Middle Cambrian clearly tie closely to central England, and the Middle Cambrian faunas of these areas contain distinctive elements also shared with the suspect terranes of southern Europe and with northwest Africa. Although all of these exotic or suspect terranes seem to have some relationship to the North African sector of Gondwana, data are insufficient to identify their original positions.

A puzzle is provided by the Cambrian of western Argentina. This is remarkably similar in both its faunas and stratigraphic details to the Cambrian of southwestern North America (Baldis and Bordonaro 1982) and is unlike the Cambrian of northern Argentina and Bolivia. Although latitudinal controls on the western Argentine faunas have been invoked (Palmer, 1974) the possibility must be considered that the Cambrian in that area is an exotic terrane tectonically "deposited" in the Andean region of South American in post-Cambrian time.

Before we can fully evaluate Cambrian biogeography and migration patterns, the problem of original locations of suspect Cambrian terranes relative to known Cambrian continents must be unequivocally resolved. Until such a reliable paleogeography is attained, most discussions of migration patterns and some aspects of biogeography will be second-order speculations of dubious lasting value.

PROGRESS IN PALEOOCEANOGRAPHY

Important clues about paleooceanographic conditions that may have influenced biogeography, marine migration and patterns of evolution in the Cambrian world are beginning to appear. These include 1) evidence for a strong thermocline as well as a distinct temperature gradient from higher to lower latitudes, 2) at least two periods of dramatic drops in sea level, and 3) evidence for strong ocean currents.

The contrast between the endemicity of the shallow-water low-latitude trilobite faunas of the carbonate-dominated inner shelves and epicontinental seas of North America, Siberia, Australia, and the suspect terranes of eastern China, and the much more cosmopolitan nature of the deeper water faunas from the ocean-facing outer shelves of these areas has been compared by Taylor and Forester (1979) with similar patterns of distribution of modern marine isopods, where the controlling factor for similar contrasts is water temperature. Isopods below the thermocline and in the cool waters at high latitutdes are much more cosmopolitan than those above the thermocline and in lower latitudes. They suggest that a strong Cambrian thermocline could explain the contrasts between the faunas in the shallower and deeper waters of the low latitude Cambrian areas, and also the more cosmopolitan nature of the shallow (cool) water faunas of the carbonate-poor facies of Baltica at high Cambrian latitudes. This important idea is central to much future thinking about biogeography.

Two periods of widespread but short-lived low stands of Cambrian sea level are now coming into focus. These are well developed in North America by stratigraphic and paleontologic evidence for mid-continent disconformities associated with seaward shifts of the major lithofacies belts across the continental margins in late Early Cambrian time, and in Late Cambrian time between the Dresbachian and Franconian stages (Palmer, 1981). The older discontinuity is also apparent between Early and Middle Cambrian rocks on Baltica. Because detection of the discontinuities that reflect these short-lived events requires detailed biostratigraphy, as well as an opportunity to place facies changes in a regional context, and good resolving power for correlations within and between regions, data to test the regional extent of these low stands of sea level is still not available for many parts of the world. However, evidence for these should be looked for at the suggested stratigraphic positions. If these low stands were global, which seems likely in the light of evidence for such low stands in younger Phanerozoic times, then they could have had an important influence on the preserved record of Cambrian evolutionary history.

Evidence for upwelling and submarine scour during parts of Cambrian time suggest the presence of strong ocean currents. Many sedimentary phospate deposits are believed to require upwelling conditions along continental margins that have favorable positions relative to wind and current patterns. The Early Cambrian phosphorites of southern Asia and the Middle Cambrian phosphorites of Queensland indicate that possibly strong upwelling was present during these parts of Cambrian time.

Recent work on the Tertiary stratigraphy of the Atlantic margin of the United States (Pinet and Popenoe, 1982) has shown that periods of strong deeper water submarine scour by the Gulf Stream are correlated with low stands of Tertiary sea level. The anomaly of late Early Cambrian and early Middle Cambrian seaward prograding clastics in the Appalachian region of North America during one of the major low stands of sea level, coupled with an absence of rocks of this age among deeper water open shelf or continental margin deposits from this same region (Palmer and James, 1980) perhaps now can be explained by submarine scour (or non-deposition).

Strong ocean currents imply the presence of a well developed wind system which further implies well-developed climatic belts and thus strong seasonality in the middle latitudes. Evidence for current and wind system vectors and for seasonality can provide important constraints on paleogeographic reconstructions.

Paleooceanographic research, as it applies to early Paleozoic time, is in its infancy but has considerable potential for contributing significant data to the search for the Cambrian world.

THE PUZZLE OF ABRUPT REGIONAL EXTINCTION

Once at the end of Early Cambrian time, once or twice in Middle Cambrian time, and three times in the Late Cambrian, the faunas of North America were abruptly and without warning nearly wiped out by events of unknown origin. The paleontologic record of the three Late Cambrian events has been studied in detail (Palmer, 1979) and, as reflected in the trilobite record, it is remarkably similar for each event: following the initial shock to the system, there is a short period of confusion, generally preserved in a meter or so of rock, during which a few local survivors are temporarily dominant, together with one or two forms of unclear ancestry. Gradually, within that short stratigraphic interval, the crisis fauna is replaced by a fauna of simple ptychoparioids of extremely low diversity that is dominated (over 90% of all specimens) by one species. This ptychoparioid fauna seems to represent a slowly evolving stock from some oceanic refugium because these simple trilobites change very little in the 5-15 million years from one crisis to the next. The low diversity fauna then gradually evolves into a new high diversity fauna before the next event. The stratigraphic reflection of these events is the abrupt replacement of a trilobite fauna of high diversity by one of low diversity and different taxonomic composition.

The interval between extinction events has been termed a biomere (Palmer, 1965) and the three youngest Cambrian biomeres are, from oldest to youngest, the Marjumiid, Pterocephaliid and Ptychaspid biomeres. Both boundaries of the Pterocephaliid biomere have been isolated within one centimeter in the House Range of western Utah. Each boundary contact lies within a bed, between similar sediments with dramatically different faunas, and is marked by a hardground.

The faunal change that marks the Mindyallan/Idamean boundary in Australia has many of the characteristics of a biomere boundary and it can be precisely correlated with the base of the Pterocephaliid biomere. Although the non-agnostid genera and species in Australia are nearly all different from those of North America, several of the same trilobite families are involved, and the agnostid trilobite species, which permit the accurate correlation, are the same. The base of the *Olenus* zone on Baltica may also reflect this biomere boundary event. In North America, this boundary is documented in detail across the entire breadth of the United States.

Whatever the cause, biomere boundaries represent events of more than local extent. The events seem to be reflected in other elements of the fauna, but except for phosphatic brachiopods, the preservation of the other elements is inadequate to document this in detail. Biomere boundary events should be searched for and documented in other Cambrian regions. They represent major, geographically widespread, evolutionary perturbations. Until we understand their origin and geographic distribution, we will be limited in our interpretations of the patterns of evolution of the Cambrian biota.

SUMMARY AND WORK FOR THE FUTURE

The theme of this symposium concerns the evolution, migration and provincialism of Cambrian organisms. Evolutionary lineages can be developed with some confidence if we restrict our studies to faunas in the shallow seas of each of the major Cambrian continents, or if we deal with cosmopolitan elements such as agnostids. Provincialism among faunas in different facies on each of the continents and some of the larger suspect terranes of Cambrian age, such as the Yangtze and North China/Korea blocks of southeastern

Asia, can be effectively studied if restricted to these regions. Integration of the information on local evolution and provincialism into a grand scenario for evolution of Cambrian organisms, in which interpretations of migration patterns would be an important part, is dependent on solution of the key problems regarding paleogeography, paleooceanography and biotic crises discussed in this paper. This challenge for the future requires the cooperative scientific efforts of a wide range of geologists and a broad-based education for each one.

REFERENCES

Baldis, B. A. J. and Bordonaro, O. L. (1982). Comparacion entre el Cambrico de la "Great Basin" Norteamericana y la Precordillera de San Juan, Argentina. Su implicancia intercontinental. Quinto Congreso Latinoamericano de Geologia, Actas 1, p. 97-108.

Ergaliev, G. Kh. and Pokrovskaya, N. V. (1977). Nizhnekembriyskie trilobity Malogo Karatau yuzhnyy Kazakhstan (Lower Cambrian trilobites of Maly Karatau-southern Kazakhstan). Nauka, Alma-Ata, 96 p.

Ergaliev, G. Kh (1980). Trilobity srednego i verkhnego Kembriya Malogo Karatau (Trilobites of the Middle and Upper Cambrian of Maly Karatau). Nauka, Alma-Ata, 179 p.

Palmer, A. R. (1965). Biomere--a new kind of biostratigraphic unit. Journal of Paleontology, v. 39, p. 149-153.

Palmer, A. R. (1974). Search for the Cambrian world. American Scientist, v. 62, p. 216-224.

Palmer, A. R. (1979). Biomere boundaries re-examined. Alcheringa, v. 3, p. 33-41.

Palmer, A. R. (1981). Subdivision of the Sauk Sequence. In: Short papers for the Second International Symposium on the Cambrian System. M. E. Taylor (ed). U.S. Geological Survey Open File Report 81-743, p. 160-162.

Palmer, A. R. and James, N. P. (1980). The Hawke Bay event: a circum-Iapetus regression near the Lower Middle Cambrian boundary. In: The Caledonides in the U.S.A., D. R. Wones (ed). Virginia Polytechnic

Institute and State University, Department of Geology, Memoir 2, p. 15-18.

Pinet, P. R. and Popenoe, Peter (1982). Blake Plateau: Control of Miocene sedimentation patterns by large-scale shift of the Gulf Stream axis. Geology, v. 10, p. 257-259.

Scotese, C. R., Bambach, R. K., Barton, Coleen, Van der Voo, Rob, and Ziegler, A. M. (1979). Paleozoic base maps. Journal of Geology, v. 87, p. 217-277.

Sun, Y. C. (1963). On the occurrence of _Xystridura_ fauna from Middle Cambrian Hainan Island and its significance. Acta Paleontologica Sinica, v. 11, p. 608-609.

Taylor, M. E. and Forester, R. M. (1979). Distributional model for marine isopod crustaceans and its bearing on early Paleozoic paleozoogeography and continental drift. Geological Society of America Bulletin, v. 90, p. 405-413.

Wolfart, Reinhard. (1963). The Cambrian System in the Near and Middle East. International Union of Geological Sciences Publication 15, 70 p.

Ziegler, A. M., Scotese, C. R., McKerrow, W. S., Johnson, M. E., and Bambach, R. K. (1979). Paleozoic paleogeography. Annual Reviews of Earth and Planetary Sciences, v. 7, p. 473-502.

THE EVOLUTION OF ECOSYSTEMS: MAIN EVENTS

PONOMARENKO, A.G.
Palaeontological Institute of the Academy of Sciences of the USSR, Moscow, USSR.

The evolution is usually considered as a process of emergence of better-adapted species of animals and plants. The evolution of ecosystems is considered only as changes of their systematic composition and the ecosystem is looked upon only as the sum of species, whose properties are quite additive and whose interaction does not lead to any special effect. The ecosystems are considered as a specific level of the organisation of life, but the specific characters of this level itself usually remain disquised. Moreover, there are no formulated concepts on the evolution of ecological systems if only their evolution is not restricted to simple change of species forming ecosystems.

Only recently has been developed a new approach to the problem of composition and function of ecological systems: a close relation between complexity and diversity of ecosystems and the mode of resource supply. It seems apparent that the synthesis of the r-and K-selection concept and the theory of ecological succession may serve as a basis for this new approach.

The main function of ecosystems is the organisation of biological circulation of energy, and the better is this organisation the more advanced is the ecological system. The main criterium of the quality of ecosystems is their ability to retain energy (as unoxided carbon) and biogenes within themselves.

The evolution of ecosystems seems to be a process of improving the above mentioned ability. The mode of functioning of the ecological succession is very similar to the historical evolution of ecological

Proceedings of the 27th International
Geological Congress, Volume 2, pp. 135–141
PALAEONTOLOGY

systems. The early steps of succession have a high productivity but the large part of organic matter is wasted from the ecosystem being buried within soil, peats and sediments. In the course of succession environments become more and more mesophilic, erosion diminishes and the rate of burying of organic matter slows down. During climax balance is established between accumulation and expenditure of organic matter while erosion practically stops. In the geological history of ecosystems we can see the same process: ecosystems become more stable, the resource redistribution is more regular and the erosion is more limited.

In ecosystems which do not provide an advanced control of resource supply r-selection is acting (McArthur, 1971; Valentine, 1971). It gives preference to species with the ability to increase rapidly the density of population during favourable, seasons and to endure very low density during unfavourable ones. These ecosystems usually have a low diversity of highly stable and low specialized species. An extreme example of such ecosystems is the system constructed only by procaryotes. Procaryotes have a very high stability practically in any physical environment where life is possible. It is noteworthy that the mass appearance of procaryotes takes place only in the environments where there are not any normal eucaryotic ecosystems. The stability of procaryotes to the physical factors of environments is much greater than that of eucaryotes. Eucaryotic systems may oppose to procaryotic ones only their better ability to regulate the level of resources in ecosystems due to the presence in them of reserve biomass of large multicellular plants and animals. The stored biomass allows to maintain the stability of active biomass. The store increases during favourable periods and is expended during unfavourable ones but active biomass remains nearly stable. An abrupt oscillation of the density of producents causes an inevitable loss of energy and biogenes from ecosystems. The density of consuments can not follow the density of producents.

Ecosystems with a low and regular resource supply have a high diversity of some K-selected species with a relatively constant and usually low density of population. The irregularity of resources supply to the ecosystem can be smoothed by reserve biomass and by additional trophic levels being introduced. These trophic levels are constructed by consuments and mainly by predators. The specialisation of producents creates in ecosystems the succession from the pioneer communities and coenophobes which can very rapidly occupy empty areas but cannot create systems with low level of resource loses. At early stages of succession the soil organic matter accumulates and the environment of system becomes more and more mesophilic, there are microclimatic, hydrographic, landscape and other types of environmental changes. Climax association in these system can successfully prevent soil erosion, diminish an amount of run-off and the output of resources nearly stops.

The proposed concept of ecosystems evolution considers the ability of ecosystem to keep resources within the system as the main criterium of their maturity. This concept seems to be easily tested by examining the dynamics of accumulation of organic carbon in sediments during geological time. However, such an attempt faces considerable difficulties. There is a decrease not in the amount of buried carbon but in the buried part of produced organic matter. When the productivity of stable systems is great, an amount of buried carbon can increase although its ratio to the whole produced organic matter decreases. Besides, a great part of organic carbon is buried not in a concentrated form, such as coal, oil, but in a dispersed state and further it will be oxidized if the sediments containing it are loose enough. As, on an average, the looseness of sediments anti-correlates with antierosional properties of vegetation, carbon buried from less advanced systems will be oxidized in greater quantity than the one buried from more advanced ecosystems. This will also obscure the suggested tendency.

The geological history of ecosystems in terms of the above formulated concepts can be described as follows. During a greater part of the history of life ecosystems were constructed by procaryotes alone. They had to have great fluctuations in biomass and density of population. A considerable part of the created biomass was buried but the primary production and biomass were rather small. Thus life existed in separate localities. The zone of life was probably limited by the shelf. The organisms could not live in the upper layers of water if the content of oxygen was very low, and the upper layers of water were sterilizated by ultraviolet rays. Euphotic zone itself in the coastal part of the seas was probably narrow. Strong erosion on continents, devoid of vegetation, prevented the development of river-lake systems. The run-off was predominantly of flanket-type, the continental sedimentation was not extensive and a greater part of pelitic material passed to marine sediments. There were no sedimentation reservoirs in the form of lakes and slowly-flowing rivers where pelitic matter and dissolved biogenes would separate. The water was muddy along the whole coast. The euphotic zone was very narrow and the thermoclyne was near to surface because muddy water absorbs sunlight in the uppermost layers of water. The decrease in the volume of euphotic zone caused a considerable decrease in productivity, while a high and sharp thermocline prevented the circulation of water and penetration of oxygen to the bottom layers of water. Thus, a great part of unoxydised biomass was buried in sediments. Such conditions promoted the formation of stromatolites which allowed algae to occupy the most preferable areas between muddy bottom layers of water and surface layers sterilized by ultraviolet rays. The high density of algae created an alkaline environment, rich in oxygen, which provided the sedimentation of not only Ca but also Mg. The inevitable irregularity of resourses supply and redistribution was one of the main factors which limited the increase of diversity and complexity of ecosystems. The rates of evolution must have been very slow under these conditions and it can be observed in the geological history. The procaryotes system existed for about two billion years.

The first important event in the evolution of ecosystems after the origin of life was the appearance of large multicellular animals and plants at the end of the Proterozoic. They created ecosystems with reserve biomass and with several trophic levels. Large multicellular algae were most important. They made the function of ecosystems much more regular and improved the supply of oxygen to the bottom layer of water. There were many large organisms in the vendian biota. With stabilization of the environment of ecosystems their diversity increased, while the density of species decreased.

The next main event was the appearence of terrestrial vegetation at the end of the Silurian. The first psilophytic associations along the sea coast quite changed the character of run-off to the ocean. A great part of pelitic material isedimented in large lagoons and the water of coastal parts of the ocean became less muddy. The process of oligotrophization of the ocean speeds up and the trophic organisation becomes more complete. Very important was the development of pellete transportation of copropel and pelitic material from the surface to the bottom. The euphotic zone became deeper and a greater part of oxygen penetrated to the bottom layers of water. With the expansion of vegetation inland erosion decreased and favourable conditions were created for the development of lakes and rivers. The extent of continental sedimentation increased while the transport of pelitic matter to the ocean decreased. The paralic basins of Paleozoic appeared in those places where the separation of pelitic matter and dissolved biogenes took place. Corals and other animals which do not survive in muddy water appeared near the coast line.

As compared with the Cainozoic vegetation, Paleozoic and Mesozoic ones have, however, a much weaker anti-erosion effect. They could form only a part of succession transformation - hydroseria. The spore and gymnosperm vegetation cannot colonize loose substrata on slopes. That is why, even with considerable predipitation in the Paleozoic and Mesozoic could exist landscapes which in appearance were similar to the present-day deserts and produced thick strata of red-coloured deposits. Still, these landscapes differed considerably from the present-day deserts. Their overgrowth by vegetation was hindered not by lack of moisture, but by a low settlement ability of vegetation. Thus sloping and loose substrata were not overgrown and were rapidly washed out, while on hard rocky substrata, which withstood erosion quite well, vegetation was considerable. In depressions, where biogenes not retained by soil were brought, vegetation was abundant and had a high productivity. It is known that present-day coniferous forests practically are equal to the decidious ones in biomass and productivity. The character of the Mesozoic landscape has been considered in detail earlier (Ponomarenko, Kalugina, 1980).

The third main event was the appearance of angiosperm plants in the Cretaceous. The angiosperms with numerous monovoltine and herbaceous forms were real revolutionaries for the Mesozoic terrestrial and water ecosystems. The ability of vegetation to prevent erosion sharply increased, the overgrowing and soil development became much more active. The accumulation of organic matter in soil and control of the run-off could diminish the transfer of biogenes to seas and continental basins bringing changes to the marine and continental water ecosystems at the end of the Cretaceous. An attempt to explain changes in Cretaceous oceanic ecosystems by the decrease in the transfer of biogenes to the ocean has been made earlier (Bramlette, 1965). Here we only suggest the mechanism of this decrease. The final formation of recent landscapes with grasslands, successionally mature forests and numerous macrophytes in continental basins took place in the Paleogene.

In conclusion, it should be emphasized that the proposed concept should not be looked upon as an absolute, one but only as a tendency. The progressive evolution of species is accompanied by the simplification and degeneration of individual forms. The same case is with the evolution of ecosystems: there constantly appear ecosystems controlled by r-selection without ability to regulate perfectly the resource supply and distribution.

It is interesting to draw attention to the analogy between Biogennetic Law and recapitulation of the history of ecosystems in succession systems. In spite of great difference between organisms and ecosystems, such tendency exists.

Unfortunately, the author cannot compare the proposed hypotheses with real data on sedimentogenesis. As shown above, a direct use of published data, such as results obtained by A.B. Ronow (1976) has proved to be impossible. Special research on different types of sediments is needed and the main intention of this paper is to draw attention to this need.

REFERENCES

McArthur R.H. 1972. Geographical ecology: patterns in the distribution of species. N.Y.

McArthur R.H., E.O. Wilson. 1967. The Theory of Island Biogeography. Princeton Univ. Press.

Bramlette M.H. 1965. Massive extinction in biota at the end of the Mesozoic time. Science, v. 148, p. 1696-1699.

Pianka E.R. 1978. Evolutionary Ecology. Harper and Row. N.Y., Hagerstown, San Francisco, London.

Ponomarenko A.G., N.S. Kalugina. 1980. The general characteristic of insects at the Manlay locality. In: Early Cretaceous lake Manlay. Trans. Joint Sov.-Ming. Paleontol. Exp., v. 13, p. 68-81 (in Russian).

Ronow A.B. 1976. Volcanoes, carbonate sedimentation, life. Geochemistry, n 8, (in Russian).

SOME ASPECTS OF STUDIES ON BIO AND PALEOGEOGRAPHY OF THE CAMBRIAN

ROZANOV, A.Yu.
(Paleontological Institute of the Academy of Sciences of the USSR, Mowcow, USSR).

Studies of the Early Cambrian biogeography carried out over many years have yielded some very important conclusions, three of which are based on the distribution of trilobites. Firstly, the presence of three major regions (realms) recognised on the basis of trilobites has been generally accepted (Hupe, 1952; Repina, 1968; 1972; Cowie, 1971; Jell, 1974; Kobayashi, 1972 etc.). secondly, principal differences in associations of trilobites depend upon the facies in which they occur. The nature of facial assemblages can vary. Palmer (1972, 1974) showed the contribution of facies and associations of trilobites around the North American craton. Repines (1972) demonstrates the dependence of associations of trilobites on their positions on "lagoon", open-marine or transitional zones of epicontinental basinal environments of the Siberian Platform. Thirdly, from our experience it became clear that in some cases there is without doubt a great difference between trilobite associations of cratons and those of "geosynclinal" regions.

Studies of the geographic aspects of Archaeocyathid distribution also allow us to outline some important trends (Zhuravleva) 1972; Rozanov, 1976 a,b; Rozanov, 1980). Firstly, it is clear that there are no great taxonomic differences between any of the known regions where archaeocyatha are known to occur. The observed taxonomic differences relate to the diversity present in each region, in a manner similar to that of recent hermatypic corals.

Proceedings of the 27th International
Geological Congress, Volume 2, pp. 143–157
PALAEONTOLOGY

Secondly, the presence of reef-like structures suggests that the majority of Early Cambrian marine basins were in lower latitude areas. Such as position would help to produce a uniform taxonomic composition of Archaeocyatha and yield more reliable intercontinental correlation (Rozanov, 1976).

Lastly, among generally uniform Archaeocyatha there are two specific associations on the basis of distribution of Anthomorpha, for example, one including southern Europe, Middle Asia and some fold regions of southern Siberia, and the other is between the Far East and those of Western Canada.

This general picture is well supplemented by data on carbonate algae which show the unity of the plant life of basins with carbonate sedimentation. In fact the main genera such as *Renalsis*, *Epiphyton*, *Girvanella* etc, are worldwide in their distribution as well as those of the Archaeocyatha. Centers of diversification of carbonate algae are connected, to considerable, extent, with those of Archaeocyatha because they both reconstruct the same bioherms.

All known data on microphytoplankton which have been mainly studied from terrigenous rocks (Palaeontology ..., 1979; Vidal, 1981 a,b; Vanguestain, 1974) also shown the unity of associations especially of Atdabanian and Botomian ages in such regions, as the East-European platform, central and North Europe, Greenland and western Canada. Unfortunately acritarchs from carbofacies are not well studied and the results cannot be interpreted properly.

A simple summary of the above mentioned conclusions does not give clear results. The great difficulties in biogeographic division of the Early Cambrian are possibly due to sharp paleogeographic changes which took place during this period of time. For example, in some regions Tommotian deposits conformably overlie more ancient ones and the structural plan does not change until the Late Tommotian - Early Atdabanian (e.g. the East - European platform, Yantzee platform and adjacent areas, Kazakhstan (Karatau), some regions in the Altae-Sayany fold belt and Mongolia). A large amount of tectonic activity during or before this time is indirectly supported by the regions where Atdabanian deposits overlie more ancient deposits than Tommotian ones (even Early Proterozoic and Archean formations). These are regions in Europe, some regions of the Eastern coast of North America, South and Central Australia. Therefore the beginning of the Cambrian saw very complicated interaction of various factors which influenced the development and migration of the fauna and flora. On one hand the Tommotian is a time of regression in many marine basins which originated in the Precambrian. On the other hand this is a time of numerous probably global changes in the geochemistry of basins. Thus, at that period we observe for the first time abundance of red limestone enriced in ferric oxides, a decrease in dolomite accumulation and the increasing role of calcium carbonates, abundant salt and phosphorite accumulation. The largest phosphorite localities in Karatau, Mongolia and S. China were formed during that period of time. Sharp changes were observed in plant and animal biotas, the most important of which was the appearance of numerous skeletized animals. At the same time stromatolites almost completely lost their stratigraphic value and were replaced by algae with calcareous shells. A great number of animals have sclerites composed of phosphatic material.

In fact the largest transgressions of the Early Cambrian started in the middle Atdabanian and reached their maximum in the Botomian, resulting in large areas being covered by Carbonate facies. It is natural that these regions could not be the most important for the migration of animal and plant organisms. The paleogeographic picture would have been simple if we had know sufficiently well the position of continental plates. An attempt to analyse the distribution of facies and "organic" remains in terms of fixist concepts will lead to large contradictions. It is now possible to take into consideration recent paleomagnetic data. Unfortunately we do not have many of the, and not all of them are trustworthy. The best results for the Early Cambrian were obtained from data on the Siberian platform, North America and Australia (Kirschvink, Barr, 1984; Kirschvink, 1978; Kirschvink, Rozanov, 1979; 1984). According to these data the Siberian platform and Australia were in the equatorial zone at least during the Tommotian and Atdabanian and North America was in middle southern latitudes. Some data were received for west Mongolia, which also suggest a position in the equatorial zone.

The reconstructions shown on fig. 8 (Rozanov, 1976) are certainly approximate schemes reflecting an idea of the existence of Palaeo-Pangea in the Early Cambrian (Rozanov, 1976). Other more moderate versions are possible (Parrish et al., 1983).

But these versions are more or less similar and differ sharply from the constructions made by Jell (1979), Gorodnitsky et al. (1978), and they allow us to link paleomagnetic, paleontological and paleoclimatic data with more confidence.

In the evalution of these various data we should bear in mind several possible pitfalls. Of great importance is the possibility that the position of some blocks could considerably change even within the Early Cambrian. Similarity, another problem with some of these past reconstructions (Gorodnitsky et al., 1978) is that many paleomagnetic poles tabulated as being of Early Cambrian age were used, but most do not have the necessary biostratigraphic control to distinguish the Tommotian from the Botomian. A third possible type of error may arise from the mis-identification of tectonic sutures between past crustal plates; not all basaltic units identified as such are indeed ophiolite fragments.

Next we must try to consider the distribution of facies and organic remains for each age of the Early Cambrian.

As was mentioned above, the Tommotian is characterized mainly by deposits of terrigenous or terrigenous-dolomite sediments in various basins. Carbonates as a rule constitute a small percent of terrigenous rocks. This type of sedimentation is typical of practically all Europe including the type Cambrian region of Wales, Central England, Normandie, Montarge Noir, Spain and certainly all of the East-European platform. The same picture is observed in Newfoundland and some other places of the eastern coast of North America. On the western coast of North America the percent of dolomites in Tommotian rocks is a little lighter (like those, for example, which contain wyattia). In Central Australia, Tommotian deposits are probably absent, but if they are present it is only as terrigenous facies (Daily, 1976, Kirschvink + Rozanov, 1984).

In some other regions, Tommotian deposits are represented mainly by dolomites. These include the upper portion of the "Li de Vin" in Marocco and the possible analogues of Tommotian deposits of the Middle East and South China, where phosphorites occur with dolomites. An analogues picture is observed in Karatau. Fold areas of the south of Siberia are characterized by a terrigenous and terrigenous-carbonate type of sedimentation (Sora formation of Kuznetsk Alatau, Anastasino formation of East Sayan, Bayangol formation of Mongolia etc.), and only in some cases there are possible units of a peculiar breccia of <u>Epiphyton</u> limestone (lower part of the Ust 'kundat formation of the Kuznetsk Alatau). Dominant normal limestone rocks are known only from the Eastern part of the Siberian platform.

Such distribution of facies could not but influence on the distribution of plant and animal associations and the position of centers of origin for some groups of skeletal fauna (Rozanov, 1976; 1980 a,b). Of great interest from this point of view are the global distributions of well studied groups. As to Archaeocyatha the situation is clear. The centres of origin and primary diversification are within a narrow belt of shallow marine bioherms which separate a large "lagoon" from the open-marine part of the basin in the Siberian platform. Other groups of organisms, such as calcarious algae, which associate themselves with Archaeocyathids in bioherm construction are also likely to appear first in this region. It is very difficult to judge if this conclusion can be drawn for such groups as gastropods, hyolithids and brachiopods, but it seems probable. The situation is much more complicated for sclerits, hyolitelmints and other problematic groups. But among all these various groups which are not sufficiently studied there are some groups which allow us to outline a series of interesting links between basins.

First of all there are the well known and morphologically distinct _Platysolenites_ and _Sabelliditides_. The deposits of the Tommotian stage in the East European platform are rich in _Platysolenites_. In Europe they also occur in the Holly Cross mountains, north of Scandinavia and Wales. But most interesting are finds of _Platysolenites_ in North America that suggest links between these territories. On the other hand _Sabelliditides_ which are abundant in the upper most part of the Precambrian and in the Lower Cambrian of the East European platform occur as well as in the north of the Siberian platform (Anabar uplift) and maybe in China and Australia. That suggests that these territories were geographically adjacent or at the least close enough for faunal migration. _Mobergella_ well known in the East European and Siberian platforms is a very important piece of evidence supporting the unity or proximity of these basins. It is more difficult to use other data, although some of them agree with this reconstruction. For example, a great number of the richest associations of molluscs, tommotiids, etc. of the Tommotian stage of the Siberian platform, Mongolia, Karatau and China are similar. Gastropods known from the East European platform (_Aldanella_ and _Anabarella_) are widespread in Siberia, molluscs of the Tommotian level in Newfoundland are also represented by _Aldanella_ and _Heraultiopegma_ genera known from Siberia. Of great interest is the development of anabaritids in Siberia, Mongolia and China as well as presence of *Protohertzina anabarica* in Siberia, Karatau and Western Canada. In all localities with Tommotian fossils there are the same taxa of hyolitelmintes.

The first half of the Atdabanian time is characterized in many regions mainly by terrigenous sedimentation. Such is the situation in the East European platform, Europe, all regions of North America as well as in Australia. However, in the second half of the Atdabanian time calcareous sediments begin to accumulate in the majority of these regions. Such are carbonates with *Bigotina* in Normandy, limestones of Home Farm formation in Central England, in W. North America the upper part of the Campito formation and the lower part of the Sekwi formation, and in Australia the lowermost parts of the Ajax limestone and their analoges, etc.

Carbonate rocks are more abundant at the beginning of the Atdabanian in Marocco, in the Altai-Sayan fold region, in Mongolia and in the Far East.

In the Middle East it is very difficult to understand the character of basins which existed at that time as the stratigraphy has not been studied well. However, strata of presumed Tommotian and Atdabian age of the Middle East are mainly dolomite-terrigenous. In China the majority of regions are characterized by the absence of Early Atdabanian deposits, or they are present mainly in dolomite facies. The second half of the Atdabanian of that region is sometimes characterized by black shales.

A small number of Early Atdabanian trilibites are known from the regions of Siberia, Northern Mongolia, Europe and Marocco. They are so peculiar in each region that they cannot be compared and the exact age is not clear. At the beginning of the Atdabanian, Archaeocyathids are found only in Siberia (in the platform and in some fold regions) and possibly in

Marocco and Spain. By the end of the Atdabanian they spread to practically all basins where calcareous sediments formed. The generic composition of Late Atdabanian Archaeocvatha is more or less uniform, and we can speak only about some specific features of associations in Marocco. Unfortunately we do not have a clear understanding of the character of distribution of other groups of organisms as they are not yet well studied. Historically, it happened that we intensively studied the Tommotian fauna in connection with the problem of Precambrian/Cambrian boundary and only now we begin to receive information on Atdabanian and later associations. There are a few small groups about which we have full information, such as the Late Atdabanian Rombocorniculum and Microdoctyon, known from Siberia, Karatau, Europe, North America.

The geographic distribution of *Volborthella* is also of great interest. It is well known in the East European platform (including the Holly Cross mountains), Scandinavia, Spitsbergen and North America. The benthic polychaete nature of Volborthella does not raise any doubts regarding the character of detailed interpretation. In this connection we need a biogeographic interpretation analogous to that of *Platsysolenites*. One should bear in mind that in the European basins *Volborthella* appears at the Early Atdabanian and in North America it appears at the Late Atdabanian-Botomian.

As was mentioned above, the Botomian is a time of maximum development of transgression and Ca-carbonate facies. At that time bituminose shale - carbonate units were deposited. Sedimentation of mainly terrigenous rocks is localised in the literal parts of basins. Whole basins are characterised by this sedimentation only in the East European platform and northern Europe.

The Early Botomian was a time of broad expansion of Archaeocyata and calcareous algae. Conversely, the end of the Botomian saw their sharp decrease in number. Only at this time we can observe a certain deferentiation in spatial distribution of Archaeocyathids. Zhuravleva (1972) recognises a number of biogeographic categories, but their rank seems to me much lower than stated.

The spread of Radiocyatha known now not only from Australia, but from the fold belts of the southern Siberia platform, North America etc., is in full agreement with the Archaeocyatha. The beginning of the Botomian age saw widespread distribution of the primitive echinodermos, and their separate plates have been found in all regions of the world. Beginning only with the Botomian age we can speak about the existence of three realms distinguished by trilobites which do not conform to the divisions made by Archaeocyatha.

At the end of the Early Cambrian during the Tojonian, the picture of the development and distribution of the fauna begins to change considerably. The beginning of the Tojonian saw an increase of trilobite associations and a sharp reduction in the number of Archaeocyatha taxa. It is more difficult to restore the paleogeographic picture of this period than that of previous ones as intercontinental correlation of the Tojonian-Amgian interval is a subject of much debate. We can only say that this time seems to be connected with the reconstruction of basins and their reduction etc. I shall give only two important facts, namely that Tojonian deposits are absent in the northern part of Europe and from almost all territory of the East European platform, and that in some areas red units were deposited like the Billy Creek formation in South Australia. The broad links which had been established between basins are evidently broken and the migration of the fauna is hampered. A small number of centres with archaeocyatha are preserved. The composition of Tojonian Archaeocyatha is very poor and monotonous but even in this case two different associations are recognized: one with Tegerocyatidae and Erbocyatidae and the other without them. The first group includes localities on the Siberian platform, the Altai-Sayan fold regions, the Eastern coast of the USA (Virginies) and Greenland. The second group consists of localities from South Australia, China, W. Newfoundland, Labrador and New Jersey, and maybe (because of shortage of material) the western part of the USA, Mexico (?), Spain and the Sinai.

Thus though we have considerable gaps in our knowledge of the Early Cambrian, the available material allows us to conclude that the majority of known paleobasins were placed in lower latitude areas. These basins determined to a considerable extent the character of development and migration of the fauna and flora, especially those connected with reef building, although trilobites migrated with their facial zones.

On the other hand, despite the moderate climate on the Earth we can outline some climatic zones including the existence of not only equatorial, but temporal belts.

Some difficulties which arise when interpreting the paleomagnetic data can be resolved by taking into account possible paleoplanktotrophic larval stages. The first attempts on this aspect have been already made (Parrish et al., 1983).

REFERENCES

Cowie J.W. Lower Cambrian faunal provinces. Faunal Provinces in Space and Time. Geo. J. Spec. Issue, 1971, N 4. Seel House Press, Liverpool, p. 31-46.

Daily B. New Date on the Basis of the Cambrian in South Australia. Izv. AN SSSR, ser. geol., 1976, N 3, p.45-52.

Gorodnitski A.M., Zonnenshain L.P., Mirlin E.G. Reconstruction of the Position of Continents in the Phanerozoic. M., Nauka, 1978, 122 p.

Hupe P. Contribution a l'etude du Cambrien inferieur et Du Pre-cambrien III de l'Anti-Atlas Marocain. - Not. Mem. Serv. geol. Maroc. 1952, t. 103, 402 p.

Jell P.A. Faunal provinces and possible planetary reconstruction of the Middle Cambrian.- J. Geol., 1974, N 3, p.319-350.

Kirschvink J.L. The Precambrian-Cambrian boundary problem: paleomagnetic directions from the Amadeus Basin, Central Australia.- Earth and Planetary Sci. Letters, 1978, v.40, p.91-100.

Kirschvink J.L., Barr T.D. The paleoposition of North America in the Early Paleozoic: New data from Caborca sequence in Sonora, Mexico. Geology, 1984 (in press).

Kirschvink J.L., Rozanov A.Yu. Magnetostratigraphy of Lower Cambrian strata from the Siberian platform: A paleomagnetic pole and a preliminary polarity time-scale. Geol. Mag., 1984.

Kobayashi T. Faunal provinces in the Early Cambrian Period. Proceed. Japan Acad., 1972, 48 N 4, p.242-247.

Palaeontology of the Upper Precambrian and Cambrian Deposits of the East-European Platform. M., Nauka. 1979, 212 p.

Palmer A.R. Problems of Cambrian biogeography.- 24 Intern. Geol. Congr., 1972, Sec. 7, p.310-315.

Palmer A.R. Search for the Cambrian world.- American Scientists, 1974, v.62, p.216-224.

Parrish J.T., Zeigler A.M., Scotese C.R., Humphreville R.G., Kirschvink J.L. Early Cambrian Paleogeography, paleoceanography and Phosphorites. In: J.H. Shergold and P.J. Cook (eds) Proterozoic and Cambrian Phosphorites Cambridge University press., 1983.

Repina L.N. Biogeography of the Early Cambrian in Siberia, According to Trilobites. M., Nauka, "Problems of Paleontology", 1968, p.46-56.

Repina L.N. Biogeography of Early Cambrian of Siberia according of trilobites. Proceed IPU. XXIII Inter. Geol. Congress, 1968, Warszawa, 1972, p.289-300.

Rozanov A.Yu. Biogeography and Early Cambrian Layers. In: "International Geological Congress. Session XXY: Communications of Soviet Geologists. M., Nauka, 1976a, p.31-36.

Rozanov A.Yu. Biogeography of the Cambrian. In: "Achievements in Science and Technology:

Stratigraphy, Palaeontology". M., VINITI 1976, v.7, p.7-24.
Rozanov A.Yu. Centres of Origin of the Cambrian Faunas. In: "Paleontology and Stratigraphy". M., Nauka, 1980, p.30-34.
Rozanov A.Yu. Centres of origin of Cambrian fauna. Congress Geologique international. Resumes Abstract. Paris, 1980, v.1, p.181.
Vanguestain M. Especes zonals d'acritarches do Cambro-Tremodaocien de Belgique et de L'Ardenne Francaise.- Rev. Palaeobot. and Palynol., 1974, bd. 18, blz. 63-82.
Vidal G. Lower Cambrian acritarch stratigraphy Scania. Sweden Geologiska Foreningens i Stockholm Forhandlingar, v.103, part 2, 1981a, p.183-192.
Vidal G. Micropaleontology and Biostratigraphy of the Lower Cambrian Sequence in Scandinavia.- Short papers 2d Inter. symp. Cambrian system, Golden, Colorado, 1981, p.232-235.
Zhuravleva I.T. Early Cambrian biogeography and Geochronology according to the Archaeocyathi. Proceed IPU, XXIII Inter. Geol. Congress, 1968, Warszawa, 1972, p.361-374.

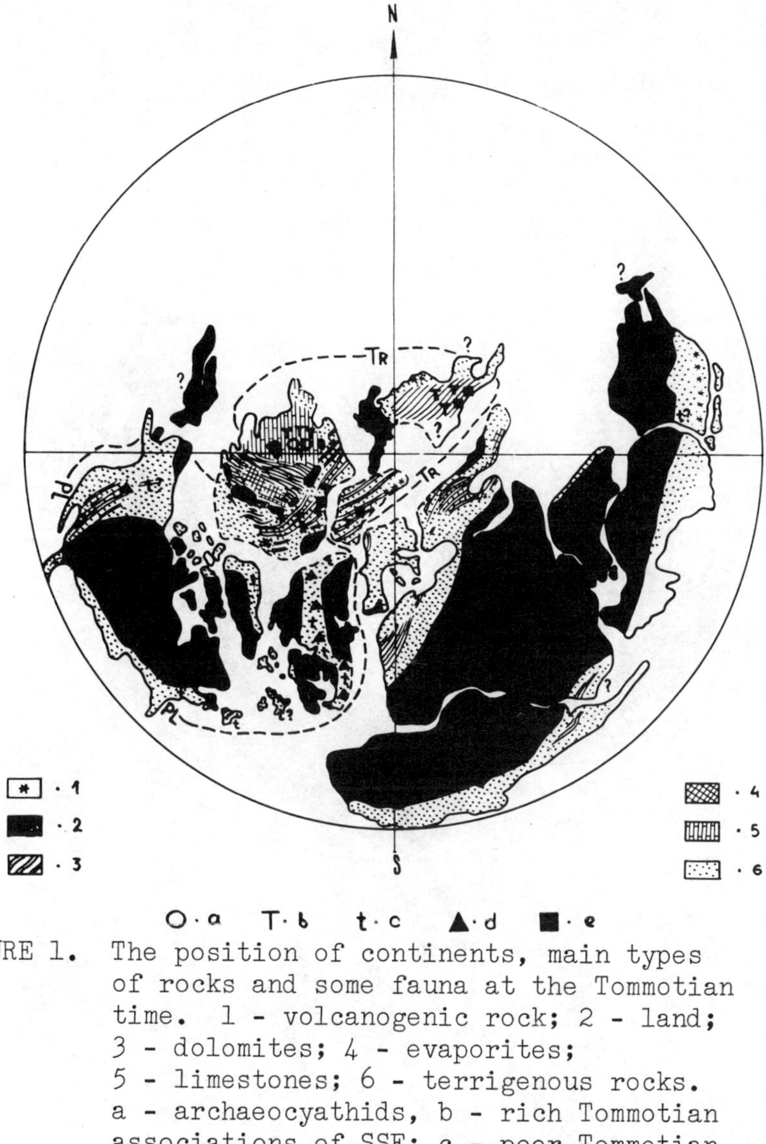

FIGURE 1. The position of continents, main types of rocks and some fauna at the Tommotian time. 1 - volcanogenic rock; 2 - land; 3 - dolomites; 4 - evaporites; 5 - limestones; 6 - terrigenous rocks. a - archaeocyathids, b - rich Tommotian associations of SSF; c - poor Tommotian associations of SSF; d - Platysolenites; e - large deposits of phosphorites. The line with the sign PL shows the area of platysolenites distribution; the line with the sign TR denotes the area of distribution of rich Tommotian fauna.

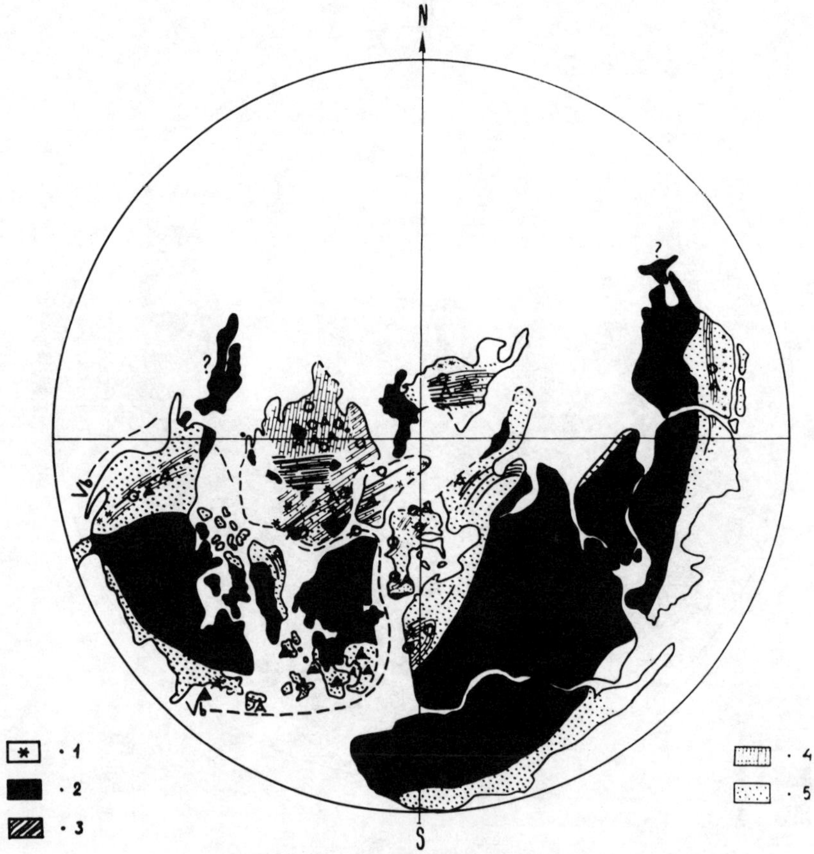

O · a A · b ▲ · c

FIGURE 2. The same at the Atdabanian time. 1 - volcanogenic rocks; 2 - land; 3 - dolomites; 4 - limestone; 5 - terrigenous rocks. a - archaeocyathids; b - the Atdabanian SSF associations; c - Volborthella. The line with the sign Vb denotes the area of Volborthella distribution.

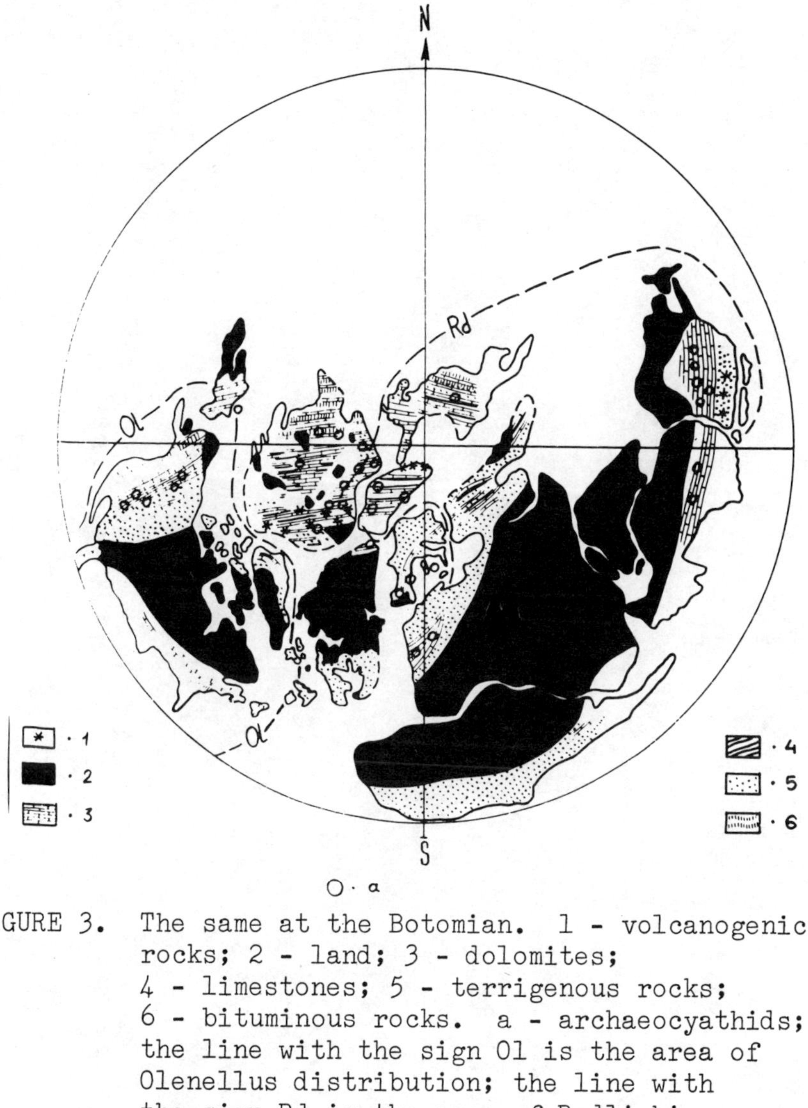

FIGURE 3. The same at the Botomian. 1 - volcanogenic rocks; 2 - land; 3 - dolomites; 4 - limestones; 5 - terrigenous rocks; 6 - bituminous rocks. a - archaeocyathids; the line with the sign Ol is the area of Olenellus distribution; the line with the sign Rd is the area of Redlichia distribution.

ORGANIC WORLD OF THE VENDIAN PERIOD

SOKOLOV, B.S., Department of Geology, Geophysics and Geochemistry of the USSR Academy of Sciences;

FEDONKIN, M.A., Paleontological Institute of the USSR Academy of Sciences, Moscow, USSR.

The studies of the Late Precambrian organic world especially intensive over last 20 years drastically changed our understanding of the Phanerozoic life. Moreover, the Phanerozoic paleobiological record which constitutes no more than 1/7 of the whole history of life on Earth has been considerably reevaluated on the basis of data on the Precambrian paleontology. The information on the Archean and Proterozoic biotas, which has been rapidly accumulated over last years makes evaluate differently the place of the Phanerozoic stage in the history of the biosphere, have a deeper insight into the role of life in the formation of the stratisphere, the hydrosphere and the atmosphere of the planet and see various ways of the development of the organic world in its interaction with abiotic phenomena.

The Vendian period terminates a long Prepaleozoi history of the organic world of Earth (Sokolov 1977). This period is to considerable extent characterized by those peculiar features which are typical of the Late Proterozoic on the whole and are not sufficiently studied by workers. On the other hand the level of the Vendian biota's differentiation seems to be similar to that of the Phanerozoic despite some significant differences of biological and ecological character. The Vendian organic world is characterized by some specific unique features which we shall try to explain below.

Proceedings of the 27th International
Geological Congress, Volume 2, pp. 159–170
PALAEONTOLOGY

The following groups of fossils have been established in the Vendian deposits; various procariotes and their life products in the form of biogenic sediments (stromatolites, microphytolites), microphytoplankton (acritarchs and other organic walled microfossils of different nature), fungi including actinomicetes, multicellular algae with non-mineralized thallom, calcareous algae with mineralized thallom, various multicellular animals, mainly soft bodied forms without mineral skeletal elements, more rarely with skeletal organic elements, first tiny organisms with mineral skeletal formations which appeared at the very end of the Vendian.

Various groups of flora including abundant microphytoplankton and macrophyte benthos flourished in the Vendian shallow water seas.

Vendotaenids covered the bottom in shallow water zones. According to the data of H.B. Gnilovskaia non-mineralized elastic thalli of Vendotaenids were multilayer strips or flattened tubes. The thalli of Vendotaenids had a tissue-like organization without vascular system, an osmatic mode of feeding and were capable of photosynthesis. Spherical reproduction organs (sporangia) have been discovered in some representatives of this group which can be compared with brown or red algae.

The systematic study of Vendotaemids (Gnilovskaia 1971, 1976, 1979; Ischenko 1979, 1982) allows us to use their remains for the biostratigraphic subdivision of Vendian deposits on the Russian Platform.

The Vendian animal world is very peculiar. There is a large majority of very specific groups dominated by Radialia which it is very difficult to place in the existing system of Metazoa.

The prevalence of Coelenterata in theVendian can be considered as a sign of the oldest age of the phyllum Eumetazoa. A wide diversity of Vendian Coelenterata recognised on the basis of symmetry reflects an earlier stage in the initial radiation of the phyllum. The most archaic are a symmetry of infinitely high order and the concentric body plan of simple forms (Nemiana, Cyclomedusa, etc.) which should be united into the class Cyclozoa. This class is characterized by concentric body plan, prevalence of polyps with thick mezoglea, non-differentiated gastral cavity and dominant sexual reproduction. Some forms (Kullingia, Ovatoscutum, Chondroplon, Dickinsonia) can be indicative of the possibility for the origin of bilateral segmented forms from the forms with concentric body plan (Fedonkin 1983), but it is difficult to see whether it was successful. The Vendian Radialia with the symmetry of variable order (Hallidaya, Bonato, Hiemalora, etc.) are characterized by a higher level of organization. These forms increase the order of symmetry in ontogenesis by increasing a number of radial channels, gonads, tentacles. The variable order of symmetry seems to be an obligatory stage in the evolution of many lines of Radialia on the way towards the forms with the fixed order of radial symmetry mainly characteristic of the Phanerozoic cnidaria.

In the Vendian this symmetry is mainly represented by 3-ray forms (Skinnera, Albumares, Tribrachidium, etc.) and by 4-ray forms (Rugoconites, Conomedusites, Ichnusina, etc.). Parallelism is recorded in the further development of Coelenterata with these types of symmetry: 3-ray forms produced angustiokreids at the beginning of the Cambrian, 4-ray forms produced conuliariids. The analysis of architectonics of the Vendian Coelenterata allows us to suppose that Radialia evolved from the forms with a symmetry of infinitely high order to the forms with fixed order, from concentric organization to predominantly radial one (Fedonkin 1983). Thus among diverse radial symmetries only few types did survive. It is noted that segmented forms are dominant while the body plans of the Vendian Bilateria are diverse with various branches of this group of animals being advanced to different extents. This phenomenon possibly indicates that bilateral symmetry and metamery are historically related and took place at the same time.

Among segmented forms thin leaf-shaped forms of the family Dickinsoniidae have the lowest level of organization. The analysis of the symmetry of representatives of this family and the comparison of their body plan with architectonics of Coelenterata, Turbelliaria and Annelidae suggest that Dickinsoniidae is an independent stem of the oldest Bilateria. The possibility of deriving the symmetry of Dickinsoniidae from that of the most archaic Radialia with an axis of rotation of infinitely high order (concentric organization) or with an axis of variable order (radial organization) can indicate that the group is extremely simple. It can't be excluded that Dickinsoniidae represent one of the stems of Bilateria which diverged from Radialia earlier than others. It is difficult to see whether this attempt was successful. But judging by the fact that Dickensoniidae, other higher Bilateria existed, Dickinsoniidae seem to have developed as a side blind alley, which ceased to exist in the Vendian, because no true finds of Dickinsoniidae are known from the Cambrian.

A large majority of bilateral and symmetrical Vendian animals are represented by flattened forms. The dominant ones are various segmented organisms, namely polymere and olygomere forms with signs of homonomous and heteronomous metamery as well as with characteristic phenomena of irregular segmentation and variable metamery. The absence or rare occurrence of animals with a worm-like form and circular or elliptical section, which are indicative of strong muscles and coelomic cavity allow us to suggest a small number of coelomates among the Vendian Bilateria. It is indirectly confirmed by specific Vendian ichnocenoses.

A large number of the Vendian Bilateria are characterized by the peculiar features which are not observed in multicellular animals of later epochs. One of the most important peculiar features is variable metamery which is characteristic of both very simple Bilateria, in particular, polymer forms with homonomous metamer and olygomer Bilateria with distinct signs of cephalization. The variable metamery can be considered as a very significant fundamental feature which reflects a necessary stage in the process of evolutionary formation of metamer in various branches of Metazoa. This type of symmetry which is not characteristic of the majority of the Phanerozoic multicellular animals possibly can be considered as one of the paleontological facts supporting the hypothesis that Articulatae are derived from radial and symmetrical organisms. But it should be emphasized once more that variable metamery of the Vendian Bilatera is indicative of the early origin of bilateral-symmetrical segmented forms from radial-symmetrical animals with an axis of rotation of uncertain order (Fedonkin, 1983).

The ecological characteristics of the Vendian period is peculiar. A group of producents in the Vendian ocean was represented by abundant phytoplankton, primarily unicellular algae and various representatives of microphyte flora of Vendotenidae. It cannot be excluded that primary consuments were larvae of invertebrates the presence of which is phytoplankton is quite possible. Secondary consuments were represented by pelagic and benthic planktonophagues and sestonophagues (meduzoids and numerous polyps among Coelenterata and other groups of Metazoa). But on the whole food chains in the Vendian world seemed to have been rather short.

Closed ecosystems suggest the presence of destructors which were possibly bacteria and to less extent, multicellular organisms (chiefly microphagues). A large quantity of organic matter accumulated on the sea-floor and was then processed by numerous representatives of vagile benthos among which, judging by trace fossils forms similar to molluscs and annelids were dominant. Widespread sapropel films in normally marine deposits of the Vendian and a high content of organic matter in sedimentary rocks can indicate that the activity of destructors in the Vendian was not optimum. This supposition is supported by paleontological data. Though trace fossils are morphologically diverse the biological processing of the sediment in the Vendian was not sufficently intensive (except for shallow water facies) because macroscopic forms of the infauna constituted only a small part of benthic communities in the Vendian seas.

The most diverse and numerous representatives of macrobenthos were Coelenterata, primarily large sedentary meduzoids as well as feather-shaped Petalonae (Charniidae and Rangeidae, Pteridiidae) and flat leaf-shaped segmented Dickinsoniidae. Microbenthos seems to have been composed of those small forms of Proarticulatae with a flattened non-differentiated body devoid of appendages which moved on the sea-floor with the help of peritaltic pedal waves of cilia. In pelagial and on the sea-floor Coelenterata were possibly dominant: various meduzoids, more rarely colonies of polyps.

To complete a short description of the Vendian animal world it should be noted that trace fossils of the Vendian invertebrates are extremely important, their study gives specific information of paleozoological and paleoecological character which sometimes cannot be obtained when studying body fossils. In the history of progressive colonization of the sea-floor the Vendian period was definitely one of the most important stages, though at that time benthic organisms occupied mainly shallow water zones of marine basins. In typical cases the Vendian ichnocenoses are characterized by a relatively low biological processing of sediment (compared with the Cambrian ichnocenoses) sub-horizontal feeding and motion traces, not deep penetration into the sediment, a small number of dwelling, burrows, resting and activity traces with appendages participating (Fedonkin, 1979, 1981).

Trace fossils from the Vendian deposits are indicative of a high organization of vagile benthos. On the sediment surface these organisms moved with the help of peristaltic changes of the whole body, as do Recent Annelidae, Sipunkulidae and other vermiculars. Among them there were sorting as well as non-sorting detritophagues which let the sediment rich in detrite and microorganisms pass through the intestine. That is shown by such traces as Planolites, Neonereites. Judging by such traces as Nenoxites, Bilinichnus, Aulichnites, other organisms moved along the surface of the sediment with the help of peristaltic waves passing through the ventral part of the body or along feet, as do recent Turbelliaria, hytous, gastropods. The peristaltic mode of motion of multicellular organisms seems to have been one of the simplest and oldest.

Traces with repeated elements of symmetry, formed by representatives of vagile benthos as a result of motions of the same type being repeated in the process of feeding are widespread in the Vendian (Fedonkin 1978, 1980). These traces are indicative of the appearance of some inherited behaviour programmes aimed at rational consumption of food resources while energy consumption is optimum. In the further history of the animal world trace fossils help us to observe the evolution of behaviour of various groups of marine benthos with progressing colonization of deep zones of the sea-floor and sediment. The evolutionary aspect of ecology which has started to develop due to studies of trace fossils will stay long the privilege of paleoichnology.

By the end of the Vendian vaginal benthos has become more active but the process of colonization of the sea-floor becomes the most rapid, especially in shallow water at the very end of the Vendian and the beginning of the Cambrian. A rich and diverse association of trace fossils appears on the Russian Platform in the deposits of the Baltic series. Many of these trace fossils persisted in later geological epochs, such as Treptichnus, Teichichnus, Phycodes, Cyrolithes, large Berganeria and others (Fedonkin 1980, 1981, Paleontology ... 1979). On the whole ichnocenoses of this level are characterized by an intensive biological processing of the sediment, diversity and morphological complexity of traces and often - by large dimentions. New forms with repeated elements of symmetry appear, the systems being not only horizontal, but vertically oriented ones. The Rowno biota is characterized by abundant Sabelliditidae, a great number of large solitary polyps (Berganeria), rich in phytoplankton, multicellular algae (flora Turasokaenia). First tubular Metazoa with non-mineralized tube (thc Nemakit-Daldyn horizon of Siberia) probably appeared at that time. The most important diversification of the organic world is observed on the Lontovaskian-Lukatian horizons of the Russian Platform and from the Sunnagian horizons and its analogues in Siberia. Mass appearance of skeletal fauna provided by the mineralization of skeletal elements in various parts of the organic world, the starting point of the Phanerozoic aeon and of the Phanerozoic evolution of the biosphere, which was the only object of classical paleontology.

D. Raup and J. Sepkoski showed in their paper that there were five mass extinctions in the history of marine multicellular animals: in the Late Ordovocian, Late Devonian, Late Permian, Late Triassic and Late Cretaceous (Raup, Sepkoski, 1982). Now we have data which allow us to state that at least one more mass extinction of many groups of Metazoa occurred at the end of the Vendian. This extinction embraced those groups which developed in the Vendian seas and represented the most characteristic elements of the Vendian fauna. These are Radialia with concentric organization and an axis of symmetry of variable order, many lines of Bilateria, especially those which represent forms with variable metamery. They ceased to exist in the Vendian as well as some colonial forms and Petalonae.

We can outline only hypotheses concerning the time causes of this first mass extinction of Metazoa. One of the possible causes of the extinction of some typical Vendian groups of Metazoa, especially numerous sedentary forms which passively caught organic detrites falling from water was the appearance of ancestors of those samll skeletal organisms the radiation of which is observed in the Lower Cambrian. Having occupied pelagial these small but numerous organisms reduce the amount of plankton falling to the bottom as they ate the main part of biomass of microscopic phyto- and zooplankton. That could lead to two consequences: the improvement of physiological, morphological, biochemical and behaviour adaptations aimed either at the optimization of the feeding processes of the elimination of passive benthic sestonophagues.

Actually both facts can be observed at the end of the Vendian and the beginning of the Cambrian. Noteworthy among behaviour adaptations are the appearance of new behaviour programmes in the process of vagile benthos, more intensive processing of the sediment by benthic animals and deep colonization (Fedonkin 1980, 1981). On the other hand, the role of active suspension feeders is increased in the zone of shallow water, primarily that of archeacyatha, sponges, and Brachiopoda in the Lower Cambrian. Suspension feeders occupied the ecological niche which had been occupied in the Vendian by large sedentary meduzoids and other organisms with simple modes of feeding. Of great interest is a sharp decrease in the number and taxonomic diversity of microphytoplankton (by 70%) in the Late Vendian (Knoll, Vidal 1982). The reasons of this phenomenon are not clear, but it cannot be excluded, that it is related to the extinction of many groups of

Metazoa in the second half of the Vendian. It is possible that a decrease in biomass of microphytoplankton promoted the selection of small forms. Being soft-bodied organisms, these forms had few chances to be preserved as fossils. Thus, poor paleontological record of the oldest Metazoa in the second half of the Vendian could be related to two various phenomena: mass extinction of large forms and a decrease in the dimensions of those ones which persisted in the Cambrian. Various trace fossils indicate that at the very end of the Vendian and the beginning of the Cambrian dimensions increased in various groups of soft-bodied Metazoa. This phenomenon has casual relationship with a new flourishing of microphytoplankton which is recorded by specialists on acritarchs. It led to the occupation of new ecological niches in the sediment of benthos, as shown by the trace fossils.

The mentioned above is indicative of unique peculiarities of the Vendian biota, its taxonomic and ecological specific character. Therefore paleontological basing of the Vendian as a terminal Precambrian system is now complete. A sharp difference between the organic world of the Vendian and that of the Cambrian, as well as probable ecogenetic completeness of many groups of the Precambrian fauna, can indicate that the Vendian - Cambrian boundary is not only the boundary between systems but the boundary between erathems.

REFERENCES

Aksenov E.M., Keller B.M., Sokolov B.S., Solontsov L.F., Shulga P.L. General Scheme of Stratigraphy of the Upper Precambrian of the Russian Platform. Izv. AN.SSSR. ser. geol. 1978, N 12, p. 17-34 (in Russian).

Fedonkin M.A. Precambrian-Cambrian ichnocoenoses of the East-European platform. Trace Fossils 2 (T.P. Crimes, J.C. Harper, ed.), Geol. Journ., 1977, spec. issue, N 9, p. 183-194.

Fedonkin M.A. Ancient trace fossils and the ways of behavioral evolution to the mud-eaters. Paleontol. Zhurnal, 1978, N 2, p. 106-112 (in Russian).
Fedonkin M.A. Fossil trace of the Precambrian Metazoa. Izv. AN. SSSR. ser. geol. 1980, N 1, p. 39-46 (in Russian).
Fedonkin M.A. Palaeoichnology of the Precambrian-Cambrian transition. In: M.E. Tailor (ed.), Short papers for the Second International Symposium on the Cambrian System. Open-File Report 81-743, 1981, p. 89-90.
Fedonkin M.A. Precambrian fauna of the Russian platform. In: M.E. Tailor (ed.), Short papers for the Second International Symposium on the Cambrian System. Open-File Report 81-743, 1981, p. 91-92.
Fedonkin M.A. White Sea biota of the Vendian. (Precambrian non-skeletal fauna of the Russian Platform north). "Nauka", Moscow, p. 100.
Fedonkin M.A. Organic World of the Vendian. "Itogi Nauki y Tekhniki". Ser. "Stratigrafia. Paleontologia", v. 12, VINITI, Moscow, 1983, p. 127 (in Russian).
Fedonkin M.A. Promorphology of Vendian Radialia as a key to understanding the early evolution of Coelenterata. Forth Internat. Symp. on Fossil Cnidaria. Abstracts. August 7-12, 1983, Washington D.C., p. 6.
Gnilovskaya M.B. Ancient aquatic plants of the Vendian of the Russian platform (Late Precambrian). Paleontol. Zhurnal, 1971, N 3, p. 101-107 (in Russian).
Gnilovskaya M.B. The Vendian Metaphyta. Bull. Centr. Rech. Explor.- Prod. Ell-Aquitaine, 1979, t.3, N. 2, p. 611-618.
Knoll A.H., Vidal G. Late Proterozoic vase-shaped microfossils from the Visingso Beds, Sweden Geol. Foren., 1980, Stockholm Forh., 102, p. 201-207.
Sokolov B.S. The Earth's organic world on the path toward Phanerozoic differentiation. Vest. Akad. Nauk SSSR. 1976, N 1, p. 126-143 (in Russian).
Sokolov B.S. The Vendian Stage in Earth History. XXIV Intern. Geol. Congr., sect. I: "Precambrian

Geology". Montreal, 1972, p. 78-84.
Sokolov B.S. The Vendian System: precambrian geobiological environment. XXVI Intern. Geol. Congr., Rep. Sov. Geol.: "Paleontology. Stratigraphy". Moscow, "Nauka", 1980, p. 9-21 (in Russian, with English summary).
Upper Precambrian and Cambrian Paleontology of the East-European Platform. (A. Urbanek, A.Yu. Rozanov, editors). Widawnictwa Geologiczne Warszawa, 1983, p. 1-158, Plates XCIV.

CLADISTIC ANALYSIS, MORPHOLOGY AND PALEONTOLOGY IN RECONSTRUCTION OF VERTEBRATE PHYLOGENY

TATARINOV, L.P.
Paleontological Institute, Academy of Sciences of the USSR, Moscow.

According to Haeckel (1866), the phylogenetics is the science which studies the ways, laws and causes of the evolution of organisms. Under such a definition, the phylogenetics as we can see makes the basic contents of the evolutionary biology, coinciding as a matter of fact with the evolutionary theory. In practice, however, the phylogenetics is concerned only with the reconstruction of organisms' relationships and the construction of the phylogenetic schemes (Tatarinov, 1976). The work of a phylogenetist is based on the analysis of the characters of similarity and distinction of the comparable organisms. The phylogenetic principles are considered to be generalizations, facilitating the estimation of these similarities and distinctions in terms of relationships (Zimmerman, 1967; von Jong, 1980). Only under such limited contents we can speak about the phylogenetics as an independent biological subject (Tatarinov, 1982).

We suggested to subdivide the main phylogenetic methods into two groups: the genotypical group, based on study of genotypes, and the phenotypical one, based on study of phenotypes. Standing apart are the methods, in which the phylogenetic conclusions are based on the results of studies not of the organisms as themselves but of their geographical and biotopical distribution and of other outward manifestations of vital activity. These methods can be conventionally combined under the name of extrasomatic; they are of subordinate significance for the phylogenetics on the whole, though turn out sometimes to be extremely important (Tatarinov, 1976).

Proceedings of the 27th International
Geological Congress, Volume 2, pp. 171–182
PALAEONTOLOGY

Of all sections of biology, the morphology - both neontological and paleontological - is still of great importance for the phenotypical phylogenetics. The contribution of paleontology made to the science is increasing constantly, and by the middle of the XX century paleontology took the leading role in the phylogenetic studies on vertebrates and most of the groups of skeletal organisms. The neontological studies give a generalized but not enough detailed picture of phylogeny though true in main features, from such results it is difficult to derive rather exact indications of the structure of ancestral organisms. Thus, for instance, without any paleontological material we could come to a right conclusion about the origin of terrestrial vertebrates from fishes, but, most likely, we would consider the dipnoans to be ancestors of the amphibians, not suspecting, e.g., that the endocranium of the closest tetrapod ancestors was divided into two units. We would, probably, consider the recent urodeles, a rather isolated by-side branch of amphibians, to be the most archaic amphibians and the closest to the real ancestors of the terrestrial vertebrates. It's likely impossible to come to a conclusion from the neontological material that massive endocranium ossification took place in the primitive amphibians. In the neontological reconstructions the phylogeny is usually a process of the harmonious transformation of the type. The phenomena of the mosaic evolution are usually underestimated by neontologists.

They are inclined to enhance the evolutionary level from which the divergent evolution of the ancestral taxon had begun, since they underestimate the possiblity of the parallel acquirement of the certain characters, diagnostic for the ancestral taxon. Therefore, the reconstructed hypothetic ancestor is supplied with many advanced characters which are lost in the true ancestor (Tatarinov, 1970, 1976).

The increasing significance of paleontology for phylogenetics is obliged not to the improved research methods (in this respect paleontology falls behind neontology but to the possiblity of the study of the heterochronous organisms including some which are close to the true ancestors and to the true phylogenetic lines. Under the prism of the paleontological analysis the phylogenetic relationships, appeared to be simple and linear, sometimes turned into the beam of slightly divergent branches, extending deeply within the limits of the ancestral taxon.

Over the last decades the cladistic methods spread more and more widely into the practice of the phylogenetic studies. This process has grown steadily only after Hennig's book was translated into English sixteen years later after the original text was published in German (Hennig, 1950, 1966). The most passionate adherents of cladism call Hennig's methodology to be revolutionary and maintain its use to be necessary in any serious studies on the phylogeny of animals and plants. The cladistic methods are widely used not only in the morphological and paleontological phylogenetic studies, but also in biochemistry, molecular biology and even biogeography. The voices of cladism's opponents (Jarvik, 1981; Panchen, 1981; Charig, 1982) are lost in the chorus of the enthusiastic adherents of cladism. The literature on cladistics increases so rapidly that within this paper any complete survey has been already impossible. The principles of the cladistic analysis, in themselves, changed essentially over the last decade. Some certain Hennig's concepts applied in the modern cladism have changed their sense, mostly, as the results of symbiose with the phylosophy of K. Popper's "critical rationalism". G. Nelson, E. Gaffney, N. Platnick from the American Museum of Natural History in New York have played a great part in the formation of "transformed" cladism, that's why the latter is named "New Yorkian".

We'll take up the short point on some cladistic aspects. First of all, we are interested in those innovations contributed by cladism into the phylogenetic studies, to which extent the modern achievements in phylogenetics are bound to these innovations and in which direction the "transformed" cladism develops.

In his "Phylogenetic Systematics" W. Hennig gives a precise formulation of the conceptual apparatus and verifies the definitions of such general terms as monophyly, paraphyly, symplesiomorphy, sinapomorphy, etc. Hennig's principles concern both phylogenetics and, to a certain extent, taxonomy. The main of them can be brought down to the following states:

1. Cladograms (phylogenetic schemes) are made according to the dichotomous principle. 2. Taxa are selected only by the vertical principle. 3. Taxa rank is determined by the sequence of their branching on the cladogram, descending from the bottom of the cladogram to the top: thus, the degree of the taxa relationship corresponds to the time of their origin. 4. All taxon diagnostical characters are subdivided into plesiomorphic (inherited, primitive) and apomorphic (derivative, progressive). 5. Taxa can be determined only by apomorphic characters. 6. The relationship criterion is synapomorphy: the sequent deviation of the various taxa can be determined on the cladogram by means of their apomorphic characters compared. 7. The pairs of taxa proceeding from one point of the cladogram form "sister groups", characterized by the most complete synapomorphy and the closest relationships. 8. One of the "sister groups" pairs preserves usually much more similarity with the ancestral taxon than the other one (the principle of deviation); both the "sister" taxa, nevertheless impart the same rank. 9. The ancestral taxon, giving a rise for two "sister" groups, disappears after this event what is determined by the requirements of the dichotomous principle in the cladogram construction.

We shan't dwell upon the merits and defects of Hennig's purely taxonomic principles, this is the scope of the classification theory rather than of the phylogenetics.

However, the significance of cladistics for the phylogenetic studies, directed towards revealing the organisms relationships, should be discussed.

The adoption of the cladistic principles brought to the essential unification of the phylogenetic studies procedure. But we don't think that cladistics, as such, would facilitate the solution of the phylogenetic problems. Thus, all the methods, used for the determination of the relationships both by cladists and adherents of the other phylogenetic schools, are based on the selection and comparison of the similar and distinctive characters and, first of all, on the analysis of the synapomorphy. The successful results of this procedure depend more on the complete and fundamental knowledge of the objects (taxa) and on the qualification of a scientist but not on the strict following the cladistic principles. As we can see, even the most experienced cladists fail in obtaining the original results unless their analysis are attended by the serious studies of a proper group of organisms. So, in the consequently cladistic analysis on vertebrates, made by Hennig himself (1983), there are no original phylogenetic discoveries; all innovations are connected with the classification, related to the consecutive application of the vertical method, combining the traditional views on the phylogeny. Exactly like this, the final estimation of one of the most original cladistic works on vertebrates, motivating the origin of tetrapods from dipnoans but not from rhipidistians, depends on how much right is the interpretation, suggested by some authors, of some morphological characters, such as, in particular, the dipnoan choana, but not on the strict application of the cladistic methods. Note, there are very serious doubts regarding the right interpretation of "choanae" in dipnoans (Jarvik, 1981).

A lot of the phylogenetic problems can be solved in different ways, depending on whether the synapomorphy of the inspecting group was developing in parallel or was inherited by all species from the common ancestor.

It is necessary to emphasize that the cladistics itself does not solve the dilemma: parallelism or monophyly (Tatarinov, 1976; Panchen, 1982). In similar cases it is necessary either to make more fundamental the analysis of the apomorphic characters or to study the new objects, on which we can trace step by step the development stages of the inspected apomorphic characters. Thus, the complete mammalian mandible found in the Upper Triassic showed the absence in them of one of the main apomorphic mammalian characters - the middle ear with three ear bones, which evidently was acquired in parallel by the monotremes and theries (Tatarinov, 1976). In general, the cladists are inclined to consider the apomorphic characters as if they are originated in one step, by one saltation without intermittent stages, paying no attention to their genesis, and this, in our opinion, is the most serious lack in the cladistic practice. This lack is insignificant while distinguishing the relationships within the low-rank taxa (genera, and partly, families), since the apomorphic roots are not deep in such cases, and its history is related to the short time intervals. However, if we are interested in the origin of the higher taxa, the apomorphy genesis becomes the main task of the phylogenetic studies, and exactly in these cases the achievements of the cladistic method are fairly insignificant.

On the whole, Hennig's principles may be considered as a deep elaborated and systematized logical concept, though, in our opinion, they cannot be characterized as a revolution in science. Their further development can be found in A. Wiley's book (Wiley, 1981). But in modern phylogenetics a lead is given not by the classic Hennigian cladistics, but by the "tranformed" one, in which as a result of complicated processes, in our view, the weak points but not the strong ones of Hennig's views were developing unilaterally. As a result, the "transformed cladistics" acquires a distinct anti-historical or, in any case, non-historical character, in consequence of which Charig calls it to be "a systematics of natural order" but not "a phylogenetic systematics" (Charig, 1982).

Anti-historicism of the "transformed cladism" can be seen in the following trends of its development:

1. The analysis of the genesis of the synapomorphic characters is actually rejected. The simple distribution of the apomorphic characters among organisms but not the morphological and historical analyses becomes the main homology criterion (Patterson, 1980). The processes of convergence and paralellelisms are under-estimated by the "transformed cladists" (Platnick, 1977; Nelson a. Platnick, 1980).

2. Over the last years there is some tendency towards the increasing underestimation or even ignoration of the significance of the paleontological material. The time scale on cladograms is often absent, correspondingly. The primacy of the recent material for phylogenetics is outlined. Sometimes it is even maintained that the fossil organisms cannot be, in general, the subject of classification (the first Løvtrup's theorem - Løvtrup, 1980).

As far as the tendencies increase, the cladograms change their forms. Formerly, the cladograms were constructed in the form of more or less symmetrically branching schemes with a time scale, immitating a phylogenetic tree, but at present, as a rule, cladograms are in the form of so-called "Hennigian comb" (Panchen, 1982). On such a "comb" all the inspecting taxa are shown in the form of consecutive branchings from the axis connecting the hypothetical common ancestor with the most specialized terminal taxon. The time scale is already absent, and the fossil forms if given are allocated on the same horizontal level as the recent ones. As a result, the cladograms look much more similar to the schemes which show only the supposing evolutionary consequence of the origin of the apomorphic characters, but not the true phylogeny. It seems to us that after having refused the attempts of the analysis of the genesis of the synapomorphic characters it became possible the construction of the cladograms which show the relationships between the three isolated species without any intermediate stages. As an example can serve the cladogram which show the probable relationships of a salmon, a lungfish and a cow (Gardiner et al., 1979). These cladograms caused Jarvik's venomous mockeries (Jarvik, 1981) and they are unlikely of a serious scientific importance if not regarding the possible methodical convenience for revealing the relationships between the taxa by their quite accidentally chosen representatives.

Rather controversial are some other modern tendencies of the cladistic evolution. Thus, the dichotomous principle, considered by Hennig as a method, turns gradually into the substantial evolutionary law (Brundin, 1963). The discussion of the problem connected with grouping of the two taxa by two branches becomes the common subject of the phylogenetic studies. Thus, discussing the phylogeny and classification of reptiles, Gaffney subdivided them into three main groups - turtles, diapsids and synapsids. He emphasizes that turtles differ from diapsids as well as from synapsids by absence of Jacobson's organ. According to Gaffney's mere allegation the absence of Jacobson's organ ought to be of great importance rather than the similar heart structure of turtles and diapsids. On this ground Gaffney concludes about the possible opposition of turtles to all other reptiles, even without any attempts of discussion the possibility of the phylogenetic development of reptiles not by two but three branches (Gaffney, 1980).

The "transformed" cladists often cite the exclusive importance of K. Popper's epistematologic principles (Platnick a. Gaffney, 1977, 1978a, 1978b; Gaffney, 1979; Platnick, 1980). There is such an influence, though, in my opinion, it is overestimated. We can't discuss in detail the Popperian concepts, but we just note, that in spite of the precise description of the conceptional apparatus, his principles as the whole, in my opinion, are one step behind, comparing with the dialectic gnoseology. The "transformed" cladists were certainly influenced by the hypothetic-deductive method of knowledge in Popper's interpretation, the principle of falsification formulated by Popper and the principle of economy adopted by Popper as well. The principle of economy is used in the phylogenetics because its adherents prefer the most "economic" hypothesis which is the most direct way of the evolution with minimum of reverses, parallelisms and convergences. But in the nature of the organic evolution there is nothing that would demand the obligatory observance of the direct and economic ways of the evolution (Tatarinov, 1976; Panchen, 1982). Panchen makes a sharp remark about the Popperian phylosophy as incompatible with the principle of economy. The anti-historicism of the "transformed" cladists has really something in common with the phylosophy of Popper who stated Darwinism to be a metaphysical theory but not a scientific one (Popper, 1976). In spite of his sympathy for the cladists (Popper, 1980). Popper, however, is not responsible for their viewpoints.

What are the reasons of such a high popularity of cladism, which is, in my opinion, disproportionate to its real achievements? It seems to me there is a certain logical error in the basis: the unification of the Hennigian procedure of the taxonomic and phylogenetic studies was apprehended by his adherents as an evidence of the objective nature of the method itself.

In our opinion, cladistics deserves such estimation only in the cases concerning the use of the consequently vertical classification principle which is itself rather controversial.

The other reason of the unusual popularity of cladistics is, probably, connected with the well-organized advertisements company which made this scientific trend the most "fashionable". Soon, cladistics will be, probably, out of fashion, though I don't think of its complete disappearance from biology as it is assumed by Charig (Charig, 1982). There are quite many strong points in cladistics which make the latter quite viable. I don't consider the "transformed" cladistics, in despite of its anti-historicism, to be a real danger for the Darwinism but, nevertheless, its decreasing influence and the refusal from the "Hennigian comb" in cladograms will be appreciated as a positive change. The process of the reducing influence of cladistics should be, as far as possible, promoted, though I think that the isolated statements of its criticists cannot render an immediate effect.

REFERENCES

Brundin L. 1968. Application of phylogenetic principles in systematics and evolutionary theory. In: Ørvig T. (Ed.), "Current problem in lower vertebrate phylogeny", Nobel Symp.

Charig Alan J. 1982. Systematics in biology: a fundamental comparison of some major schools of thought.- In: Joysey K.A. a. Friday A.E. (Eds.), "Problems of phylogenetic reconstructions", Syst. Assoc. Spec. Vol., No. 21, p.21, p.363-440. Academic Press, London and New York.

Gaffney E.S. 1979. An introduction to the logic of phylogeny reconstruction.- In: Cracraft J. a. Eldredge N. (Eds.) "Phylogenetic analysis and palaeontology", p. 79-111. Columbia University Press, New York.

Gaffney E.S. 1980. Phylogenetic relationships of the major groups of Amniots.- In: Panchen A.L. (Ed.), "The terrestrial environment and the origin of land vertebrates". Syst. Assoc. Spec. Vol. No. 15, p. 595-610. Academic Press, London and New York.

Gardiner B.C. 1980. Tetrapod ancestry: a reappraisal. -In: Panchen A.L. (Ed.), "The terrestrial environment and the origin of land vertebrates". Syst. Assoc. Spec. Vol., No. 15, p. 177-185. Academic Press, London and New York.

Gardiner B.C., Janvier P., Patterson C., Forey P.L., Greenwood P.H., Miles R.S. a. Jefferies R.P.S. 1979. The salmon, the lungfish and the cow: a reply.- Nature (Lond.), vol. 277, p. 175-176.

Haeckel E. 1866. Generelle Morphologie der Organismen.- Berlin, 567 S.

Hennig W. 1950. Grundzuge eine Theorie der phylogenetishen Systematik.- Berlin, Deutscher Zentralverlag, 232 S.

Hennig W. 1966. Phylogenetic systematics. Urbana, University Illinois Press, 263 p.

Hennig W. 1983. Stammgeschichte der Chordaten.- Verlag Paul Parey, Hamburg und Berlin, 149 S.

Jarvik E. Lungfishes, tetrapods, paleontology, and plesiomorphy. Syst. Zool., vol. 30, No. 3, p. 378-384.

De Jong. R. 1980. Some ttols fro evolutionary and phylogenetic studies.- Z. zool. Syst. und Evolutionforsch., Bd. 18, Nr. 1, S. 1) 23.

Løvtrup S. 1975. On phylogenetic classification. - Acta zool. Cracov., vol. 20, p. 499-523.

Nelson G.I. a. Platnick N.I. 1980. A vicariance approach to historical biogeography.- BioSciences, vol. 30, p. 339-343.

Panchen A.L. 1982. The use of parsimony in testing phylogenetic hypotheses.- Zool. J. Linn. Soc., vol. 74, p. 305-328.

Patterson C. Cladistics. - Biologist, vol. 27, p. 234-240.

Platnick N.I. 1977. Parallelism in phylogeny reconstruction. - Syst. Zool., vol. 26, p. 93-96.

Platnick N.I. 1980. Phylosophy and transformation of cladistics.- Syst. Zool., vol. 28, p. 537-546.

Platnick N.I. a. Gaffney E.S. 1978a. Evolutionary biology: a Popperian perspective. - Syst. Zool., vol. 27, p. 137-141.

Platnick N.I. a. Gaffney E.S. 1978b. Systematics and Popperian paradigms.- Syst. Zool., vol. 27, p. 381-388.

Popper K.R. 1976. Unended quest. An intellectual autobiography. - Fontana - Collins, Glasgow.

Popper K.R. 1980. In: "Letters: Evolution". - New Scient., vol. 87, p. 611.

Rosen D.E., Forey P.L., Gardiner B.C. a. Patterson C. 1981. Lungfishes, tetrapods, paleontology, and plesiomorphy._ Bull. Amer. Mus. Natur. Hist., vol. 167, No. 4. p. 163-375.

Schmalhausen I.I. 1964. Origin of land vertebrates. Moscow, "Nauka", 272 p. (in Russian).

Severtsov A.N. 1939. Morphological laws of evolution. Mowcow-leningrad, Izdat. Akad. Nauk SSSR, 610 p. (in Russian).

Sokal R.R. a. Sneath P.H. 1963. The principles of numerical taxonomy.- W.H. Freemen and Co., San Francisco.

Tatarinov L.P. 1970. Some problems of the phylogenetic studies on the lower tetrapods. In: "Materials on the evolution of vertebrates". Moscow, "Nauka", p. 8-29 (in Russian).

Tatarinov L.P. 1976. Morphological evolution of theriodonts and general problems of phylogenetics, Moscow, "Nauka", 258 p. (in Russian).

Tatarinov L.P. 1982. Morphology, paleontology and phylogenetics. In: Problems of evolution of animal morphology. Moscow, "Nauka", p. 20-31. (in Russian).

Wiley E.O. 1981. Phylogenetics. The theory and practices of phylogenetic systematics. - John Wiley and Son, New York, 439 p.

Zimmerman W. 1967. Methoden der Evolutionswissenschaft (Phylogenetik). - In: Heberer G. (ed.), "Die Evolution der Organismen", Bd. 1, S. 542-611, Stuttgart.

GLOBAL EVENTS AND EVOLUTION

OTTO H. WALLISER

Geologisch-Paläontologisches Institut, Goldschmidt-Str.3, D-3400 Göttingen.

ABSTRACT

Global bio-events are caused by geological processes, which themselves may be triggered by initial, predominantly abiotic events or profound environmental changes. The bio-event -- often coinciding with a litho-event -- is characterized by an extinction/radiation sequence. The radiation can be explained as an acceleration of the evolutionary rate, caused by a lowering of the selectional stress. Thereby the molecular processes, i.e. quality and quantity of mutations and recombinations, do not require any change. In this biological respect the gradualistic point of view, based on Darwin's theory, is still valid.

INTRODUCTION

Since the 26th International Geological Congress at Paris, 1980, the International Palaeontological Association (IPA) is concerned with a Research Programme about "Fundamental biotic changes in earth history (global events and evolution)".

In the meantime publications, colloquia, and conferences about global events in the biosphere as well as in the lithosphere have greatly increased. The fact that long periods with "normal", relatively slow evolution alternate with relatively short lasting but important changes in certain parts of tbe biosphere, are well known since the beginning of our geosciences. These events gave rise for a subdivision of the Phanerozoic. Yet, the biological

Proceedings of the 27th International
Geological Congress, Volume 2, pp. 183–192
PALAEONTOLOGY

interpretation has changed manifold in the course of time. In the pre-Darwin time, the events have been interpreted as catastrophes, according to the scriptural Deluge (e.g. G. de Cuvier). Then, as a reaction to Darwin's theory, gradualism dominated all evolutionary considerations, denying sudden changes, which have been explained by a lack of information and/or of geological record (e.g. Ch. Lyell). In the early decades of this century, into the 40ies, fundamental biotic changes ("Faunen-Schnitte") have been discussed again. Major genetic transformations in an early ontogenetic phase, orthogenesis, degeneration, and geological events have been used as an explanation for radiations and extinctions (e.g. O.H. Schindewolf). At the same time also more gradualistic views (e.g. G.G. Simpson) have been discussed and subsequently preferred. However, since a few years the pendulum again swung back to a kind of neocatastrophism. The old discussion was revived, mostly under the labels of new terms and/or definitions, such as "punctuated equilibrium" or "macroevolution versus microevolution", and, most recently, enriched by bolides. The purpose of the present paper is to ask for the interactions of geological or physicochemical processes with the biological mechanisms, which lead to those fundamental biotic changes.

WHAT IS AN EVENT?

A global event should be globally detectable. The event may be restricted to only a certain part within the whole range of e.g. palaeogeographical, sedimentological, facial, ecological, or biological realms. The time span of an event itself is remarkably shorter than the intervening times of stable conditions or slow changes and developments, respectively.

It should be distinguished between initial, i.e. causative events, and final, i.e. consecutive events. Initial

FIGURE 1 (next page). Flow chart about possible causes and interactions of geological processes, which may lead to global events.

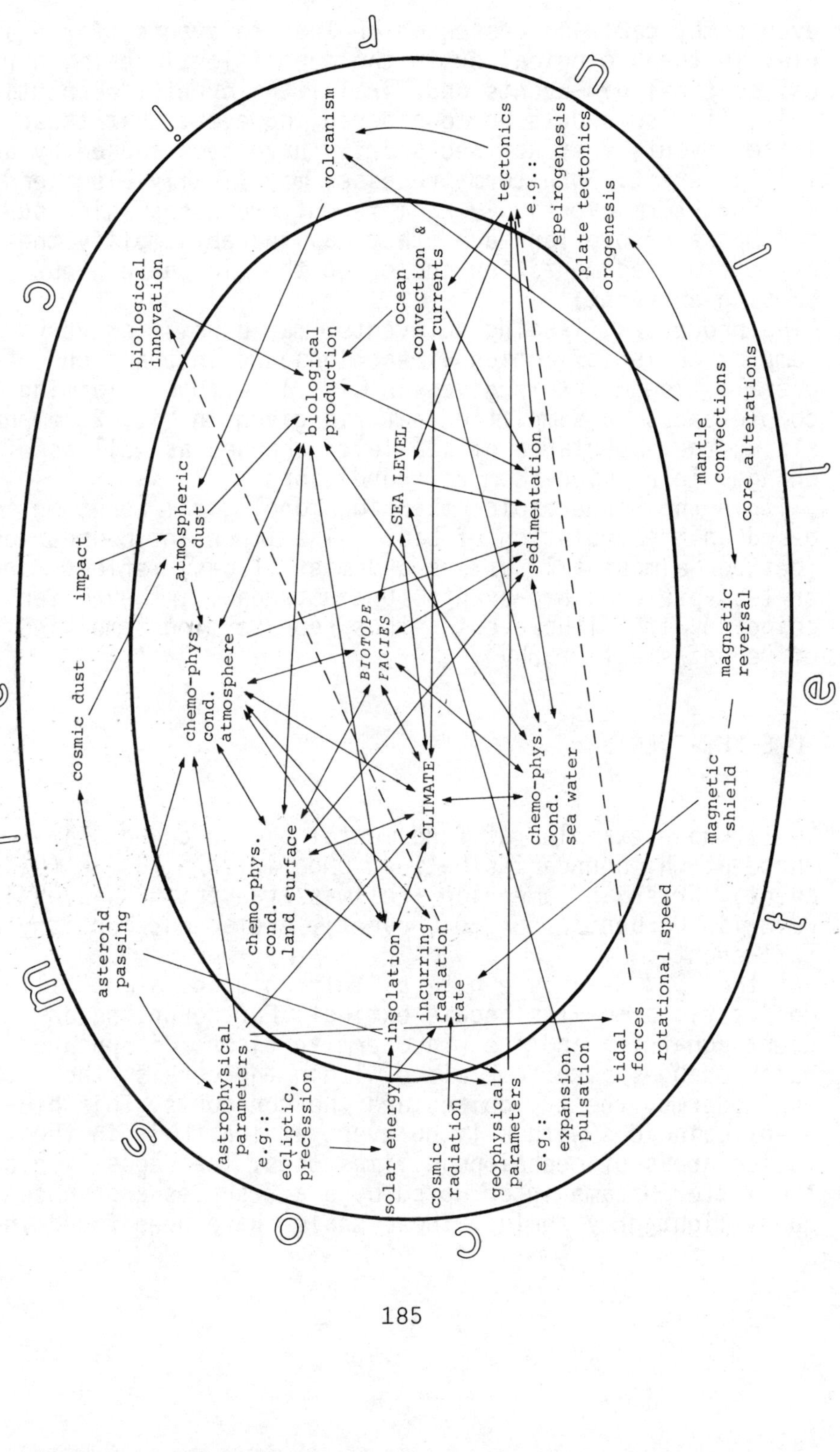

cosmic
telluric
cosmic dust
impact
asteroid passing
astrophysical parameters
e.g.: ecliptic, precession
solar energy
cosmic radiation
geophysical parameters
e.g.: expansion, pulsation
tidal forces
rotational speed
insolation
incurring radiation rate
chemo-phys. cond. land surface
chemo-phys. cond. atmosphere
atmospheric dust
biological innovation
biological production
BIOTOPE FACIES
CLIMATE
SEA LEVEL
ocean convection & currents
volcanism
tectonics
e.g.: epeirogenesis plate tectonics orogenesis
sedimentation
chemo-phys. cond. sea water
mantle convections
core alterations
magnetic reversal
magnetic shield

events may cause processes which lead to events, for example, in the biological or in the depositional realm, here called final bio-events and final litho-events, respectively. It also should be considered, however, that those latter events must not necessarily have been caused by an initial event: long-term processes may finally also lead to short-term events. Even different processes which superimpose or overlap each other may thereby amplify the effects on sedimentation and/or on the biosphere, thus causing an event.

The processes, leading to events, may consist of very complex chains of actions, reactions and interactions. An overall flow chart is given in Fig. 1, without claiming completeness. A simplified part is given in Fig. 2, emphasizing the importance of sea-level changes as well as of changes in physico-chemical conditions.

Within the Phanerozoic, the biochronological zonation is based on the evolution of biota. The important boundaries (between almost all systems and most stages) reflect more or less profound bio-events. Understandably, they often coincide with litho-events, thus representing some kind of "event-stratigraphy".

SOME KEY-EVENTS

Well known examples of fundamental biotic changes are those at the boundaries between Ordovician/Silurian (O/S-event), Frasnian/Famennian (Kellwasser-event = KW-event), Devonian/Carboniferous (D/C-event), Cretaceous/Tertiary (C/T-event).

At the D/C-event all clymenid and almost all goniatitid ammonites became extinct. Furtheron the ancient agnathian and the placodermian fishes disappeared. Major changes also took place within the corals, the echinoderms, the trilobites and the conodonts. This bio-event coincides with a litho-event at least within the facies areas of cephalopode limestones. The deposition of the latter became interrupted by black shales and subsequent light-grey shales. These shales have been found in

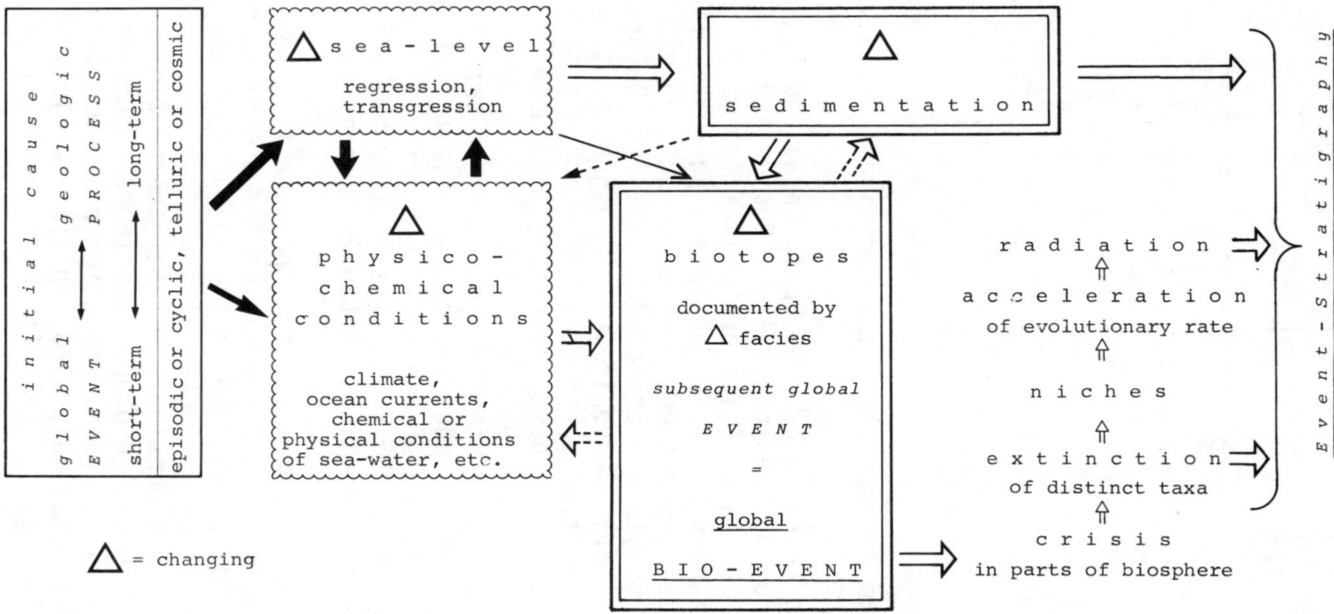

FIGURE 2. Simplified diagrammatic flow chart, demonstrating influences and interactions in the course of geological processes which are started by an initial cause and lead to final events.

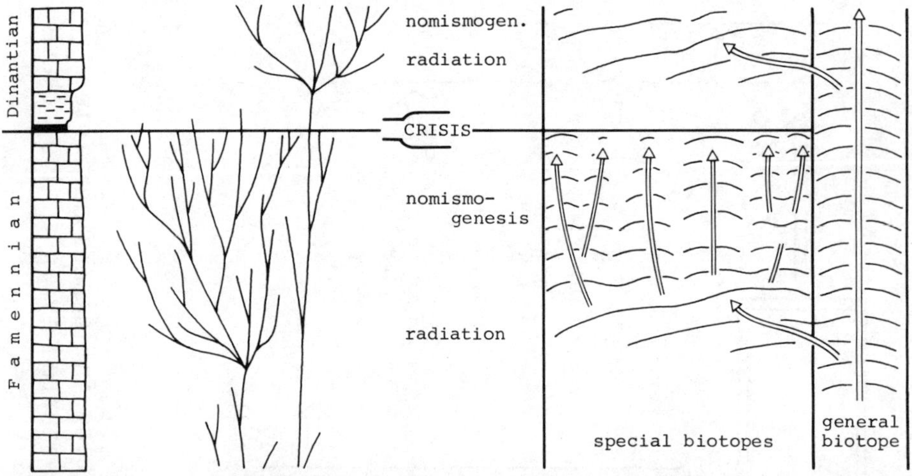

FIGURE 3. Diagrammatic sketch of the ammonite's evolution around the D/C-event.

all corresponding sections between South China and Southwest Canada, including the Iran, all of Europe and North Africa. Whereas the black shales are only a few mm to cm in thickness, the grey shales ("Hangenberg shales") often measure a few meters, sometimes only cm or dm. The event is contemporaneous with the change from the late Devonian regression to the early Carboniferous transgression.

Transgressions are often connected with black shales, i.e. with anoxic events. Those especially influence benthic organisms. At the D/C-boundary, thus, those goniatites survived, which lived in the free water column ("general biotope" of Fig. 3). But those specialized forms which had been controlled by the sea floor, as e.g. the clymenids, disappeared. The reason was most probably the extinction of one or more links of the food chain (e.g. certain algae).

With respect to the ammonoids, the important features are the following: (1) the precondition for the first radiation of clymenids within the Famennian was the innovation of a new construction (internal siphuncle). (2) The radiation implies an expansion from the general biotope into neighbouring special biotopes and thereby certain specialisations. (3) The rapid radiation phase was followed by a time of relatively slow, "normal" evolution (nomismogenesis in Fig. 3). (4) After the crisis, the free positions in the reestablished and in the newly generated biotopes made possible a renewed expansion, i.e. radiation, of the ammonoids.

The coupling of extinction and subsequent radiation is a general feature which occurs in connection with all known bio-events. This, for example, occurred within the graptolites as a result of the O/S-event; during the KW-event the same happened within the goniatites, and in connection with the extinction of the dinosaurs during the C/T-event the radiation of the mammals has to be considered.

The C/T-bio-event is suitable for the recognition of the complicated network of processes and interactions. As a result of the final Cretaceous regression the extremely widespread epicontinental seas vanished completely. This caused the extinction of all organisms which had been especially adapted to the specific conditions of epicontinental seas. Even the biotope of those dinosaurs which had been adapted to the formerly widespread costal swamps, became extremely reduced, if not almost completely lost. On the other side new terrestrial biotopes came into existence. This trend was certainly intensified by a change of the climate, which had also been affected by the regression through a considerable change of the albedo. A change in the ocean currents due to the regression and/or the contemporaneous widening of the Atlantic Ocean may have amplified the climatic change. With both changes, in biotopes and climate, the preconditions for the radiation of angiosperms had been fulfilled, i.e. a change of the primary food supply for terrestrial animals. I dare to claim that all those processes -- caused by the regression -- are sufficient to explain most features of the C/T-bio-event. In contrast to that, it would be very difficult to explain the event only by an impact of a cosmic body. If this impact really happened, it only amplified or accelerated the existing processes.

As mentioned above, global events are often connected with sea-level changes: e.g. Middle Cambrian (transgression, connected with a long lasting insertion of dark shales), O/S (caused by the end-Ordovician glaciation), D/C (see above), Permian/Triassic (regression/transgression), C/T (see above).

In contrast to the above listed examples, the K W - e v e n t is not with certainty connected with a sea-level change, but with an anoxic event. The latter is globally documented by the Kellwasser-horizon (called after a locality in the Harz Mountains and recognized for decades representing a major event near the Frasnian/Famennian boundary). The KW-horizon comprises one to several, but often two black layers. Those contain, in the realm of cephalopode limestones, layers or lenses of black limestone, whereas, in basinal locations, only black shales are present. The lower KW-horizon is within the upper part of the Lower gigas-conodont Zone, the upper KW-horizon within the Uppermost gigas-Zone.

The KW-event factually caused a fundamental bio-event: major extinctions and changes took place in many groups, such as brachiopods, trilobites, ammonoids, placoderms etc.. Yet, the most serious disturbance to the biosphere was the world-wide elimination of the biohermal reefs.

BIOLOGICAL MECHANISMS

The deciding features with respect to the evolution are: long lasting periods of "normal" evolution alternate with relatively short times in which a phase of extinctions is followed, with some retardation, by a phase of radiations. In times of "normal" evolution, which I would like to call nomismogenesis, the evolution usually proceeds relatively slow, i.e. the evolutionary rate is relatively low. However, it is important to be aware of the fact that during the nomismogenesis new constructional schemes may develop (e.g. Bactritoidea becoming Ammonoidea by coiling [as an example for relatively quick development] ; transition from reptiles to mammals [as an example for long evolutionary processes]). These biological innovations may lead to radiations which are either independent of the extinction/radiation sequence (e.g. first radiation of ammonoids) or caused by it (e.g. Tertiary radiation of mammals).

The innovation during the nomismogenesis do not need special mechanisms, such as saltation, neomorphoses, ma-

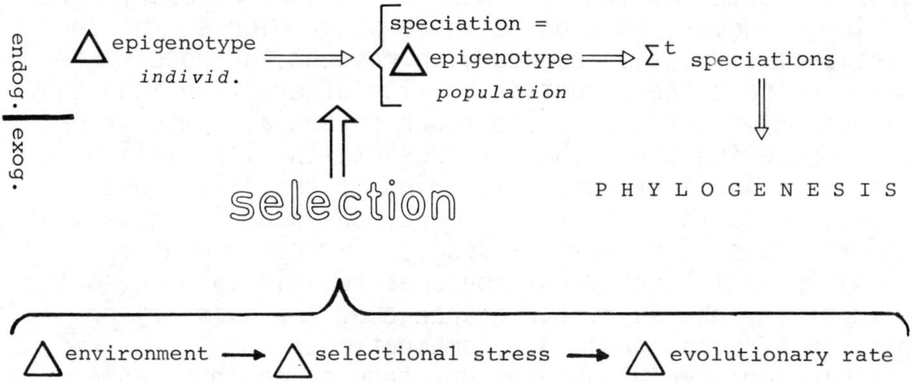

FIGURE 4. Selection is the decisive parameter, through which exogenic processes influence endogenic ones. A change (triangle) of the selectional stress influences the evolutionary rate.

croevolution or whatever else. There are no mechanisms necessary other than those which lead to "normal", i.e. billion times exercised occurrence of new species, to the speciation. In this connection it may only be noticed that the question, whether speciation occurs sympatrically or peri-petrically or allopatrically or any one of those, is independent of the question about "gradualism: yes or no". Gradualism, in the author's opinion, means that an alteration of the genetic code, which leads to an evolution, i.e. to an alteration of a certain population, is carried out in very small steps ("microevolution") according to the findings of molecular biology.

There still remains the question about the biological mechanisms which are responsible for the peculiar features of radiations. As already pointed out, radiations are often coupled with preceeding extinctions or with the creating of new biotopes. In both cases niches have been produced. Under this term I comprise free, i.e. unoccupied places in preexisting or new ecosystems.

In an undisturbed ecosystem the selectional stress is relatively high. This means that only a few of the new occurring mutations and recombinations successfully contribute to the evolution. Therefore the evolutionary rate is low. However, as soon as there occur niches, the selectional stress is reduced with respect to those forms which possess the potential ability of evolving into the direction of the niche. The fewer potential competitors for conquering the niche are present, the less will be the selectional stress. A lowering of the selectional stress means that relatively more new occurring mutations and recombinations may be used for further evolution. Thus, the evolutionary rate becomes relatively high, but without any necessity for a change of the quantity or quality of mutations and recombinations.

Conclusion: Evolution does not take place in an uniform manner, but it is episodically fluctuating with respect to the evolutionary rate. Evolution takes place gradualistically with respect to the molecular biological mechanisms. The episodic acceleration of the evolutionary rate is caused by a decrease of the selection pressure and the latter itself by extraordinary imbalances of the ecosystems. The crises in the ecosystems are produced by global events. In so far global events play an important role in phylogenesis. They have their effect in accelerating, in pushing forward the evolution.

EARLY PHANEROZOIC DISTRIBUTION PATTERNS OF SOME MAJOR GROUPS OF SESSILE ORGANISMS

WEBBY, B. D.
Department of Geology and Geophysics, University of Sydney, N.S.W. 2006, Australia.

ABSTRACT

The Early Phanerozoic fossil record includes a number of groups of organisms which are well skeletonized, non-shelly and sessile: the archaeocyaths, some sponges, the bryozoans, stromatoporoids, tabulate corals and rugose corals. The temporal and spatial distribution patterns of each of these groups is reviewed briefly. Most of the major groups apparently made their earliest appearances in platform areas - archaeocyaths in the Siberian Platform, stromatoporoids and tabulate corals in the North American Platform and bryozoans in Baltoscandia. However the clathrodictyid stromatoporoids first appear in the island-arc setting off the Australian sector of Gondwana and the heliolitine corals first diversify there. Dispersal patterns are difficult to assess but two groups, the archaeocyaths and the clathrodictyids, spread remarkably slowly around the world. Sponges appear to have their origins as poorly skeletonized forms of deeper slope environs but also dispersed into platform areas. Bryozoans seem to have originated in the temperate zone of Baltoscandia but then immigrated to equatorial areas, and even to near polar areas. The review concludes with a discussion of the significance of the major changes in distribution of well skeletonized sessile benthos through Cambrian and Ordovician time, especially in the 'reef' community.

INTRODUCTION

In briefly reviewing the patterns of early appearances of well-skeletonized, sessile, non-shelly organisms, I intend to focus on temporal and spatial aspects of

Proceedings of the 27th International
Geological Congress, Volume 2, pp. 193–208
PALAEONTOLOGY

distribution - spatial because some of the groups exhibit meaningful patterns of dispersal from inferred 'evolutionary' centres. However it must be emphasized that what we presently take to be 'real' patterns may prove to be illusory when more adequate world-wide documentation of the various fossil groups becomes available. One needs to be fully aware of the shortcomings of present documentation, and of such examples as Burrett and Richardson's (1978) survey of Cambrian trilobites. They were able to show a pronounced direct correlation between fluctuations in total generic diversity through the Cambrian and monographic effort. Moreover, as Sheehan (1977) has shown, the number of species described for each period reflects only how much effort has been expended on systematic descriptions. Relative to the rest of the Phanerozoic, species diversity is low because of smaller numbers of active workers describing Cambrian and Ordovician fossils.

This review focusses on the distribution patterns of six groups of organisms - archaeocyaths, sponges, bryozoans, stromatoporoids, tabulate and rugose corals - and attempts to explain some significant aspects of these distributions (Figure 1). I have used tripartite divisions of the Cambrian and the Ordovician. In the Early Cambrian, the Siberian stages, Tommotian, Atdabanian, Lenian and Elankian have been employed. The Mindyallan and Idamean represent the two earliest stages of the Australian Late Cambrian. The Tremadoc and Arenig are regarded as Early Ordovician, the Llanvirn, Llandeilo and Caradoc (to the top of *Dicranograptus clingani* zone) as Middle Ordovician, and the latest Caradoc and the Ashgill as Late Ordovician. The Australian graptolite zone of *Dicranogr. hians kirki* correlates with either the British zone of *D. clingani* or zone of *Pleurogr. linearis*.

ARCHAEOCYATHS

These are the most diverse, sessile skeletonized Early Cambrian organisms (Figure 1). They achieved a very widespread geographical spread (Figure 2A), and their calcareous skeletons contributed significantly to the

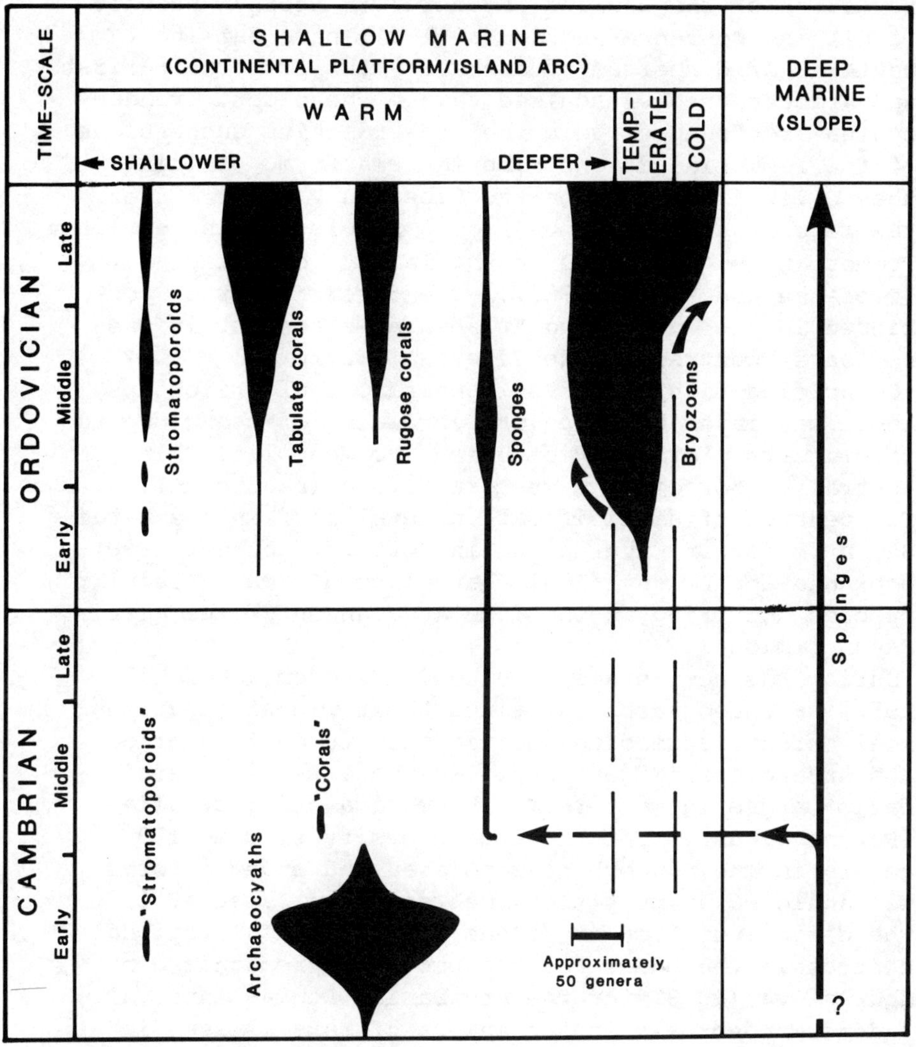

FIGURE 1. Diagrammatic representation of distribution of major sessile non-shelly animals in the chief environmental regimes of the Cambrian and Ordovician. Note that in showing the spread of bryozoans in cold-water environments it has been necessary to exaggerate the number of Late Ordovician genera beyond the 87 recorded.

formation of the earliest Phanerozoic reefs. Rozanov (1981) and Debrenne and James (1981) have outlined patterns of dispersal of archaeocyaths from their first appearances as skeletonized taxa. The oldest archaeocyaths are found only in the pre-trilobite successions of the Tommotian of the Siberian Platform. Initially they diversified only on the Siberian Platform; for the regular archaeocyaths from 6 species in the earliest Tommotian to 49 species in the latest Tommotian (Debrenne and Prieur 1981). The diversification continued in the succeeding Atdabanian stage but it was no longer centred on the Siberian Platform.

Coinciding with the first appearances of trilobites, there was major geographical spreading of the group to other parts of the Soviet Union, to Mongolia, China, Australia, Morocco and western Europe (Figure 2A). Two centres of diversification developed, in the Altai Sayan region of Siberia and in Morocco (Rozanov 1980). Debrenne and Prieur (1981) have listed some 87 regular species occurring in the late Atdabanian of the Altai Sayan region.

During the next stage, the Lenian (Botomian of Hill 1972), archaeocyaths attained their widest geographical extent, spreading additionally to North America and Antarctica (Figure 2A). New centres of diversification arose in western North America and Australia (Rozanov 1980). The highest diversity of some 180 genera including both cosmopolitan and endemic forms was achieved at this time (Debrenne and James 1981). The differentiation of faunas in the various regional centres is the basis for the provinces recognized by Zhuravleva (1968) for the Atdabanian-Lenian interval. A drastic decrease in the number of taxa is seen in the succeeding Elankian (Lenian of Hill 1972), with an equally marked reduction in the spread of forms. In a few areas, for example North America, Siberia and Australia, the remaining stocks evolved in isolation prior to their final extinction at the beginning of the early Middle Cambrian.

SPONGES

While sponge spicules are frequently identified in Cambrian and Ordovician successions, articulated skeletons are much less commonly found. Of the three major groups of sponges, the Demospongiae produced the only really well skeletonized siliceous forms during the Cambrian. These were the lithistides, and well preserved representatives are first known from the early Middle Cambrian (Kruse 1983). But it was not until the Early Ordovician that this group became especially common. One lithistide family, the Anthaspidellidae, contributed significantly to Early Ordovician algal/sponge reefs (Toomey and Nitecki 1979) and diversified to some 17 genera by early Middle Ordovician times (Finks 1970). Sponges of the other two major groups had less rigid, thinner-walled skeletons. These included the siliceous hexactinellides of *Protospongia* type and the calcareous heteractinellids like *Eiffelia*. Both had appeared by Middle Cambrian times. The sponge faunas are not well enough known to assess zoogeographical relationships. However they clearly occupied two quite different environmental settings, the shallow water carbonates and reefs, and deeper water 'slope' habitats (Figure 1). There are many records of the muddy 'slope' deposits exhibiting discrete sponge spicules, less commonly of prolific faunas of entire specimens (Conway Morris and Whittington 1979). Indeed it would seem that the siliceous sponges at least originated in the deeper 'slope' environs of Palmer's (1979a) 'outer detrital belt' during Late Precambrian or Early Cambrian times, and then colonized the shallower carbonate platform areas in the Middle Cambrian after the demise of the archaeocyaths.

BRYOZOANS

Although Bassler's (1911) records of bryozoans from the Tremadoc of Estonia are no longer well substantiated (Larwood and Taylor 1979), Baltoscandia remains the most likely area of 'origin' for the group (Figure 2B), and

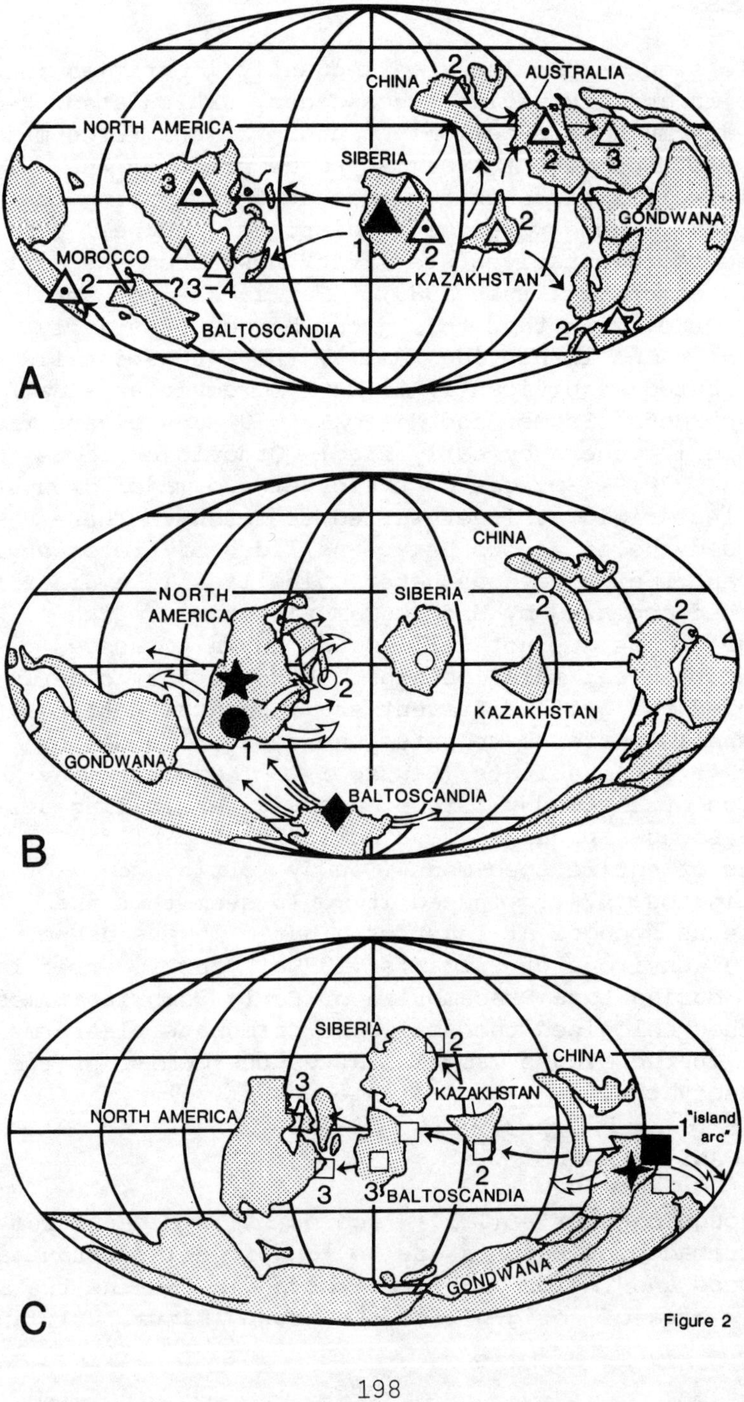

A
CHINA
AUSTRALIA
NORTH AMERICA
SIBERIA
GONDWANA
MOROCCO
KAZAKHSTAN
BALTOSCANDIA
?3-4
B
CHINA
NORTH AMERICA
SIBERIA
KAZAKHSTAN
GONDWANA
BALTOSCANDIA
C
SIBERIA
CHINA
KAZAKHSTAN
NORTH AMERICA
"island arc"
BALTOSCANDIA
GONDWANA

Figure 2

certainly it has the earliest diverse faunas (Spjeldnaes 1982). The bryozoans presumably colonized this platform area while it was in the temperate zone (intermediate to high latitudes according to Bergström and Noltimier 1982), like the area off the south coast of Australia still occupied by a great abundance of bryozoans today (Wass et al. 1970). The groups seems to have spread fairly rapidly to low latitudes, for the earliest bryozoan occurrence in North America is of Arenig age (McLeod 1978). But the bryozoans did not become a prolific, widely distributed group until the Middle Ordovician (Figure 1). Some 91 genera have been recorded (Larwood and Taylor 1979). These include members of Middle Ordovician patch reefs and shelf-edge reef complexes (Walker and Ferrigno 1973), possibly forms which lived in relatively deeper (and cooler) water positions of the reefs.

Bryozoans also became well adapted to cooler-water environments (Mediterranean province of Havlíček 1974) in Late Ordovician times (Figure 1). Some even

---

◀FIGURE 2. Cambrian and Ordovician base maps of Scotese, Ziegler and others showing first appearances, and geographical spread of certain groups of organisms.
A. Distribution of archaeocyaths in the Early Cambrian: centre of origin (solid triangle), diversification centres (large triangles with centred dot) and other first appearances (small open triangles). Timing: 1 = Tommotian, 2 = Atdabanian, 3 = Lenian, 4 = Elankian.
B. Possible Early Ordovician evolutionary centres for stromatoporoids (solid circle), bryozoans (diamond) and tabulate corals (5-pointed star). Timing of stromatoporoid appearances: 1 = centre of origin in Early Ordovician, 2 = main dispersal (open circles) in early Middle Ordovician.
C. Distribution of clathrodictyid stromatoporoids in Middle Late Ordovician: centre of origin (solid square) and other first appearances (open squares). Timing: 1 = middle Caradoc, 2 = late Caradoc - early Ashgill, 3 = middle-late Ashgill. Note also earliest appearance of heliolitine coral faunas (4-pointed star).

occupied positions close to the Ordovician pole - for example, in Morocco a rich bryozoan fauna is preserved in 'islands' of an otherwise glaciated area (Spjeldnaes 1981). An analogous relationship is seen in the Permian of Tasmania (Rao 1981). The bryozoan-dominated Berridale Limestone has associated glacial dropstones, and formed in cold waters no more than 10° of latitude from the pole.

STROMATOPOROIDS

Stromatoporoid-like structures in the Early Cambrian of the Altai Sayan region of Siberia have been interpreted as aberrant archaeocyaths, the oldest stromatoporoids and members of a small independent group (Webby 1979). There is no record of analogous Middle or Late Cambrian structures but in the Early Ordovician of the western United States (Toomey and Nitecki 1979) the problematical reef-building organism *Pulchrilamina* has the generalized growth form, latilaminae, cysts and pillars of typical later Ordovician labechiids. The main diversification of labechiids occurred in the Middle Ordovician (Llandeilo-early Caradoc), essentially worldwide but in the equatorial belt (Webby 1980). They were among the most important frame builders in the development of Ordovician reefs, growing in the shallowest and most turbulent zone.

A second important group, the clathrodictyids, first appeared in an island arc setting - shallow waters of a small offshore volcanic 'high' - in eastern Australia during middle Caradoc times (Figure 2C), and from this probable evolutionary centre dispersed slowly worldwide (Webby 1980). It took from middle-late Caradoc time to the end of the Ordovician for the group to spread from Australia through Central Asia to Baltoscandia, and then to cross the Iapetus Ocean to eastern North America (Anticosti Island). Clathrodictyds achieved the status of a major cosmopolitan group only in the Silurian. Possibly the clathrodictyids arose independently and do not share a common ancestry with labechiids.

CORALS

Of the many coral-like structures described from Cambrian successions only a few have been interpreted as skeletonized zoantharians, and none seems to have been on a direct line of descent to tabulates or rugosans (Scrutton 1979; Hill 1981). All these attempts at producing skeletons appear to have occurred in geographical isolation and to have been short-lived.

Hill (1981) has provided a highly authoritative review of the early evolution of tabulates. The major diversification occurred in the Middle Ordovician, with representatives of all the main groups (chaetitides, tetradiids, sarcinulides, favositides, heliolitides and auloporides). Hill (1981) has listed some 33 Middle Ordovician tabulate genera rising to some 78 genera by Late Ordovician times. Possibly the North American platform was the main evolutionary centre (Figure 2B) for the Middle Ordovician fauna (Kaljo and Klaamann 1973). The first tabulate (*Lichenaria*) appeared in the Early Ordovician of North America (Pratt and James 1982) and by Llandeilo (Chazyan) times chaetitides and sarcinulides had established themselves in frame-building and binding roles in reef growth (Pitcher 1971). For most genera the essentially worldwide 'equatorial' spread seems to have been relatively rapid but some constituents of the fauna continued to maintain more geographically restricted distributions through Middle and Late Ordovician times (Kaljo and Klaamann 1973).

One group of forms, however, does not appear to have had its early development in North America. These are the coenosteoid representatives of the Order Heliolitida (including suborder Heliolitina and some members of suborder Halysitina). The earliest diverse heliolitine faunas appear in an island-arc setting - shallow waters of a small volcanic offshore 'high' - in the Middle Ordovician (middle Caradoc equivalents) of eastern Australia (Figure 2C). The faunas which are presently under study are more diverse than originally indicated by Hill (1957), and occur in horizons immediately predating the zone of *Dicranogr. hians kirki*. The first coenosteoid halysitines (Webby and Semeniuk 1969) also made their initial appearance in the same island-arc setting, in slightly higher beds of the succession. This suggests that the eastern Australian 'island-arc' was the diversification centre for these coenosteoid faunas. The two groups only achieved a world-wide 'equatorial' distribution in later times, the Heliolitina during the Late Ordovician, and the coenosteoid halysitines in the Silurian. The possibility exists that these coenosteoid forms arose independently of other tabulates.

Whether the rugose corals represent an offshoot of tabulates (Flower 1961) or evolved independently from a separate soft-bodied zoantharian stock (Scrutton 1979), they remained the least conspicuous of the two groups in the Ordovician - their generic diversity rose from 18 in the Middle Ordovician to 42 in the Late Ordovician (Hill 1981). Although not yet adequately documented, there appears to be a crude, two-fold zoogeographic differentiation of faunas into (1) those including both solitary and compound which seem to occupy, in association with stromatoporoids and certain tabulates (e.g. tetradiides), the warm equatorial zone and (2) those composed only of solitary rugosans, which may represent associations of the more temperate zone (or occupy relatively deeper, cooler waters of the equatorial zone). The Australian faunas for instance have abundant occurrences of compound *Favistina* and *Palaeophyllum* and are therefore of warm equatorial aspect, whereas successions in Baltoscandia are dominated by solitary rugosans supporting suggestions that the region mainly lay in the temperate zone (Jaanusson 1972). Only after the North European plate drifted into low latitudes during the Ashgill did compound rugosans appear (Spjeldnaes 1964). Solitary rugosans have even been recorded from the Late Ordovician succession of Brittany, from an area assigned to the 'cooler' Mediterranean province (Spjeldnaes 1981).

Again there seems to be significant early appearances in eastern Australia, emphasising the role of the 'island arc' as an evolutionary centre for new groups. *Hillophyllum* which appears in the Middle Ordovician (early-middle Caradoc) is the earliest rugosan to exhibit well-defined monacanthine septa (Webby 1971), and *Rhabdelasma* which first occurs in the Late Ordovician (middle Ashgill) is the earliest to show rhabdacanthine septa (McLean and Webby 1976).

DISCUSSION

The record of earliest appearances of some fossil groups has been interpreted in terms of possible origins and dispersal, and seems to show meaningful patterns. Centres of origin of major groups (phylum to subclass) like the archaeocyaths, tabulate corals, stromatoporoids and bryozoans seem to be in platforms areas, whereas taxonomic divisions of lower rank (order to family), for example, clathrodictyid stromatoporoids and heliolitine corals appear to have originated or first diversified in island arcs. Spjeldnaes (1981) has noted that while some areas around tectonically unstable volcanic islands may be ideal places for organisms to evolve they may disappear completely when the volcanic islands become a part of major tectonically deformed and metamorphosed orogenic belts. Therefore such areas may be difficult to identify. So far only areas in the Ordovician of the Appalachian-Caledonide orogen (Neuman 1972) and in the Tasmanides of eastern Australia (Webby 1980) have been recognised.

The most striking feature of the overall pattern of distribution of Early Phanerozoic well skeletonized, non-shelly, sessile benthos, is the poor Middle-Late Cambrian record (Figure 1). Echinoderms also show a marked decline in the Late Cambrian (Sprinkle 1980) and molluscs and brachiopods remain in 'suspended animation' until near the end of the Cambrian. Given the impoverished nature of the sessile benthos in Middle-Late Cambrian shallow marine habitats, it seems unlikely that competition from, for example, sponges, was responsible for the extinction of archaeocyaths. It appears equally improbable that a major regression at the end of the Early Cambrian was the cause (Brasier 1982), since a lowering of sea level should displace not remove the continental and island shelf margin habitats. Two other possibilities seem more worthy of consideration. Firstly, large predators like arthropod *Anomalocaris* are found in relatively shallow water deposits of the Early Cambrian (Briggs and Mount 1982)). If *Anomalocaris* occupied a position near the top of the food chain, other smaller, more lowly, predators may also have existed, some adapted to feed on the archaeocyaths.

Secondly, the archaeocyathan extinction may mark the first of the four (or five) faunal crises which characterize the 40 million-year interval to the end of the Cambrian. These faunal crises are best documented in populations of trilobites occupying shallow platform areas of the near equatorial belts in western North America, Australia and China (Öpik 1966; Palmer 1979b), where there was massive turnover of genera and species. Of the 81 species in the uppermost Mindyallan Stage of Australia, none survived into the overlying lowest zone of the Idamean Stage (Öpik 1966). This latter zone has only 10 species, but the number continues to build up in succeeding zones to a total of 100 species, as the shallow platform was repopulated by new stocks from the adjacent ocean. Öpik (1966) has proposed a sudden transitory increase in solar heat, and Palmer (1979b) a rapid cooling of the shallow epeiric seas. Perhaps both views can be more-or-less reconciled if the cause was extraterrestrial. Of the many hypotheses, that of a cometary impact, as outlined by Hsu (1980) to explain the major extinction at the end of the Cretaceous, is perhaps the most appealing. Heated, cyanide-contaminated waters would ascend to the surface and converge in equatorial regions causing the greatest havoc. Middle latitudes largely escaped the catastrophe and eventually provided cooler-water stocks for re-population of the equatorial belt.

The dramatic increase in variety of skeletonized animals, including the major radiations of echinoderms, brachiopods, molluscs and bryozoans at the beginning of the Ordovician followed the emergence of a notable degree of provincialism and the return of a reasonable level of environmental stability, especially to platform areas. In consequence a large number of habitats became available for colonization. The most narrowly adapted and environmentally sensitive areas were in the reef, and these, after largely being devoid of skeletonized animals from the end of Early Cambrian times, again became occupied by a variety of forms. By Middle Ordovician times a rather complex community structure had emerged, with representatives of algae, stromatoporoids, bryozoans, corals, sponges and other organisms, probably for the first time, entirely filling the ecological space of the reef habitat in shallow equatorial waters.

REFERENCES

Bassler, R. S. (1911). The early Palaeozoic Bryozoa of the Baltic provinces. Bull. U.S. natn Mus. 77, 1-382.

Bergström, S. M. and Noltimier, H. C. (1982). Latitudinal positions of Ordovician continental plates based on paleomagnetic evidence. Paleont. Contr. Univ. Oslo 280, 8.

Brasier, M. D. (1982). Sea-level changes, facies changes and the late Precambrian-early Cambrian evolutionary explosion. Precambrian Research 17, 105-123.

Briggs, D. E. G. and Mount, J. D. (1982). The occurrence of the giant arthropod *Anomalocaris* in the lower Cambrian of southern California, and the overall distribution of the genus. J. Paleont. 56, 1112-1118.

Burrett, C. F. and Richardson, R. G. (1978). Cambrian trilobite diversity related to cratonic flooding. Nature 272, 717-719.

Conway Morris, S. and Whittington, H. B. (1979). The animals of the Burgess Shale. Sci. Amer. 241, 122-133.

Debrenne F. and James, N. P. (1981). Reef-associated archaeocyaths from the Lower Cambrian of Labrador and Newfoundland. Palaeontology 24, 343-378.

Debrenne, F. and Prieur, A. (1981). Computerization of regular archaeocyathan files. Int. Symp. Concept. Meth. Paleo. Barcelona, 313-317.

Finks, R. M. (1970). The evolution and ecologic history of sponges during Palaeozoic times. Symp. Zool. Soc. Lond. 25, 3-22.

Flower, R. H. (1961). Montoya and related colonial

corals. Mem. Bur. Min. Res. New Mexico 7, 1-97.
Havliček, V. (1974). Some problems of the Ordovician in the Mediterranean region. Ustřed. Ustav. Geol., Věstnik 49, 343-348.
Hill, D. (1957). Ordovician corals from New South Wales. J. Proc. R. Soc. N.S.W. 91, 97-107.
Hill, D. (1972). Archaeocyatha. In: Treatise on Invertebrate Paleontology, Part E, C. Teichert (ed.). Geol. Soc. Am. and Univ. Kansas, pp. E1-158.
Hill, D. (1981). Rugosa and Tabulata. In: Treatise on Invertebrate Paleontology, Part F, Coelenterata, Supplement 1, (in 2 vols), C. Teichert (ed.). Geol. Soc. Am. and Univ. Kansas, pp. F1-762.
Hsu, K. J. (1980). Terrestrial catastrophe caused by cometary impact at the end of Cretaceous. Nature 285, 201-203.
Jaanusson, V. (1972). Aspects of carbonate sedimentation in the Ordovician of Baltoscandia. Lethaia 6, 11-34.
Kaljo, D. and Klaamann, E. (1973). Ordovician and Silurian corals. In: Atlas of Palaeobiogeography, A. Hallam (ed.). Elsevier, Amsterdam, pp. 37-45.
Kruse, P. D. (1983). Middle Cambrian '*Archaeocyathus*' from the Georgina Basin is an anthaspidellid sponge. Alcheringa 7, 49-58.
Larwood, G. P. and Taylor, P. D. (1979). Early structural and ecological diversification in the Bryozoa. In: The Origin of Major Invertebrate Groups, M. R. House (ed.). Academic Press, London, pp. 209-234.
McLean, R. A. and Webby, B. D. (1976). Upper Ordovician rugose corals of central New South Wales. Proc. Linn. Soc. N.S.W. 100, 231-244.
McLeod, J. D. (1978). The oldest bryozoans: new evidence from the Early Ordovician. Science 200, 771-773.
Neuman, R. B. (1972). Brachiopods of Early Ordovician volcanic islands. Proc. 24th Int. Geol. Congr., Sect. 7, 297-302.
Öpik, A. A. (1966). The early Upper Cambrian crisis and its correlation. J. Proc. R. Soc. N.S.W. 100, 9-14.
Palmer, A. R. (1979a). Cambrian. In: Treatise on Invertebrate Paleontology - Part A, Introduction, R. A. Robison and C. Teichert (eds). Geol. Soc. Am. and Univ. Kansas, pp. A119-135.

Palmer, A. R. (1979b). Biomere boundaries re-examined. Alcheringa 3, 33-41.
Pitcher, M. (1971). Middle Ordovician reef assemblages. Proc. N. Am. Paleont. Convention (Chicago), Part J, 1341-1357.
Pratt, B. R. and James, N. P. (1982). Cryptalgal-metazoan bioherms of early Ordovician age in the St. George Group, western Newfoundland. Sedimentology 29, 543-569.
Rao, C. P. (1981). Geochemical differences between tropical (Ordovician) and subpolar (Permian) carbonates, Tasmania, Australia. Geology 9, 205-209.
Rozanov, A. Y. (1980). Tsentry prioskhozhdeniya kembriyskikh faun. In: Mezhdunarodnyy geologicheskiy kongvess XXVI sessiya. Doklady sovetskikh geologov. Paleont. Stratigr. B. S. Sokolov and A. I. Zhamoida (eds). Nauka, Moskva, pp. 30-34.
Scrutton, C. T. (1979). Early fossil cnidarians. In: The origin of major invertebrate groups. M. R. House (ed.). Academic Press, London, pp. 161-207.
Sheehan, P. M. (1977). Species diversity in the Phanerozoic: a reflection of labor by systematists? Paleobiology 3, 325-328.
Spjeldnaes, N. (1981). Lower Palaeozoic palaeoclimatology. In: Lower Palaeozoic of the Middle East, Eastern and Southern Africa and Antarctica. C. H. Holland (ed.). Wiley, Chichester, pp. 199-256.
Spjeldnaes, N. (1982). Ordovician bryozoan faunas. Paleont. Contr. Univ. Oslo 280, 49.
Sprinkle, J. (1980). Early Diversification. In: Echinoderms: notes for a short course, T. W. Broadhead and J. A. Waters (eds). Univ. Tenn. Dept. Geol. Sci., Studies 3, 86-93.
Toomey, D. F. and Nitecki, M. H. (1979). Organic buildups in the Lower Ordovician (Canadian) of Texas and Oklahoma. Fieldiana. Geol. N. Ser. 2, 1-181.
Walker, K. R. and Ferrigno, K. F. (1973). Major Middle Ordovician reef tract in east Tennessee. Am. J. Sci. 273A, 294-325.
Wass, R. E., Conolly, J. R. and MacIntyre, R. J. (1970). Bryozoan carbonate sand continuous along southern Australia. Mar. Geol. 9, 63-73.

Webby, B. D. (1971). The new Ordovician genus *Hillophyllum* and the early history of rugose corals with acanthine septa. Lethaia 4, 153-168.
Webby, B. D. (1979). The Ordovician stromatoporoids. Proc. Linn. Soc. N.S.W. 103, 83-121.
Webby, B. D. (1980). Biogeography of Ordovician stromatoporoids. Palaeogeog. Palaeoclimatol. Palaeoecol. 32, 1-19.

THE ECOSYSTEMS OF THE JURASSIC AND EARLY CRETACEOUS BASINS IN THE NORTH OF SIBERIEA

ZAKHAROV, V.A., SHURYGIN, B.N.
Institute of Geology and Geophysics, Siberian Branch of the USSR Academy of Sciences, Novosibirsk, USSR.

Jurassic and Neocomian seas bordering Siberian platform in west, north and east were epicontinental basins of the boreal type. The largest of them, West Siberian, was existing during Late Jurassic and Neocomian, covering the plate of the same name. The area of the marine surface of this basin reached 2 million square kilometers, and the depth exceeded 500 metres, approximately. The North Siberian basin of the temperate depth occypying during all Jurassic and Neocomian the Enisey-Lena depression drawned from west to east, bordered it in the north-west. The least shallow Lena-Viljuy basin bordered Siberian platform in the east in the Early and Middle Jurassic. The area of its surface during that time reduced to the north (Fig. 1, 2).

The situation of the mentioned sedimentary basins in the same climatic belt, the permanent connection with a central Arctic sea and vicinity of the large massive of land (the source of the transportation of sedimentary material) determined similarity in the main characteristics of their ecosystems: the wide distribution of terrigenous sedimentation with predominance of sands, aleurolites and claysand with sharp subordination of carbonate constituent (concentration of carbonates in form of concretions and concretionary bands was at the step of early diagenesis) and development of biota of the boreal type: the prevailing of the boreal molluscs (Bivalvia, Cephalopoda, Gastropoda) and of the Protozoa (Foraminifera and Radiolaria) and complete absence of corals, Miditerranian molluscs (Rudista, Nerinea) and Protozoa (Tintinnida), very scarce Echinoidea and of other lower Boreal

Proceedings of the 27th International
Geological Congress, Volume 2, pp. 209–222
PALAEONTOLOGY

and Tethys groups of fauna.

The area of the surface and dimensions of the North-Siberian seas were steadily increasing during Jurassic and reached their maximum at the end of Jurassic and the beginning of Neocomian. Then the basins gradually reduced. With that the aquatories shifted from east to west, that was connected with partial closing of the geosynclinal seas in Early and Middle Jurassic in the eastern sector of Arctic and with expansion of the boreal transgression in the Late Hauterivian on the west territory of the former North-Siberian seas. The continental conditions had set in.

The terrigenous sedimentation prevailed over chemogenic one in the sedimentary basins to a considerable extent. Desintegrated material was brought to the North-Siberian seas mostly from south (Middle-Siberian and North-Kazakhstanian lands) and only partly from north (Taimyr island and others). The high standing of the land (in the early Lias, in part of Middle Jurassic, in Oxfordian) was accompanied by transport of coarse-detrital material; at the periods of peneplanation (early Toarcian, Volgian age - Berriasian) chemical wearing became more intensive. At that time the clay bituminous deposits (Kiterbut and Bagenov horizonts) were formed (Palaeogeography ..., 1983).

The absolute depthes of the North-Siberian seas are not determined exactly. The relative depthes occured within the limits of sublittoral and partly of pseudo-abyssal. In the Early and Middle Jurassic the depthes of the upper and middle sublittoral (Lena-Viljuy, Khatanga basins) were prevailing. The conditions of the lower sublittoral of that time were presented only on the limited territory of the Khatanga water area. In the Late Jurassic and at the beginning of the Early Cretaceous the regime of lower sublittoral had become settled on the substantial area, and in the volgian age and at the beginning of Berriasian in the central part of the West Siberian sea on the area about 1 million square kilometres the pseudo-abyssals spread to the axial parts of the Enisey-Lena sea-strait. The large amount of Early Hauterivian and partly Valanginian deposits formed under the conditions of extreme shoal water (Zakharov, Yudovni, 1974).

Salinity of the water of the North-Siberian sea hardly substantially deflected from that of oceanic at that time, because they were freely connected with central Arctic of oceanic type basin. It is possible however that in the littoral regions and in the shallow bays the water was freshened due to the river run-off from the surrounding land masses. The supply of the fresh water especially rose at the periods of humidisation of climate (for example in Early Toarcian and Hauterivian). The paleothermometry data for Neocomian of Khatanga depression conform this suggestion. For lagoon-marine conditions salinity, determined by oyster shells, was 30%, for the littoral-marine of shallow water - 32% and for the open moderate deep-sea-water - 35% (Zakharov, Radostev, 1975). There are ideas in the literature about the general freshening of water in the Arctic basin in Jurassic based on the results of the analysis of absorved complexes of the alcaline kations in the clays of sea genesis and on other geochemical indices: Li/Ga, B. The general reducing of salinity could probably take place at the periods of the paleogeographic isolation of the Arctic basin, for example, in the Middle Jurassic. However, for the greater period of Early, the whole of Late Jurassic and for Early Neocomian either paleothermometry or paleoecology evidence do not confirm this suggestion (Zakharov, Saks, 1980).

Temperature was the most variable factor during Jurassic and Neocomian, because the territory of Siberia was in the zone of sesonal temperate-warm (Subtropical) climate: semiarid in the south and semihumid in the north (Golbert et all, 1968). The northern geographic pole was situated in the limits of Chukotka-Alaska region. The seas on it's territory were more cold. The average annual temperatures of the superficial water and of the littoral-shallow water areas in the north of Siberia formed 11-24°C according to the paleothermometry data. Repeated fluctuations of the average annual temperatures values had taken place in that interval (Zakharov, Saks, 1980). The most essential cooling is supposed for the most period of the Middle Jurassic and for the middle of Neocomian. In the Early and Late Jurassic the temperatures maximums are marked. The bottom waters of the deep part of shelf always was more cold than the surface waters, and at the periods of the cooling in Arctic their temperature was as low as 0°C (Kaplan, 1978; Kemper, Schmitz, 1981).

In the north-Siberian seas, like in the modern seas, there were different currents. The wave and probably tidal currents are recorded very well by the distinctivenesses of the burial of shells. Directed bottom and surface currents are reconstructed on base of the features of the fauna geographic distribution and by data of the water paleothermometry. The presence of the convectional circulation in the deep pseudo-abyssal hollows is probably due to thermal stratification of waters. In the Early and Middle Jurassic the principal surface currents were directed from east to west, in the Late Jurassic and Neocomian these were directed from west to east. The deep currents had contrary direction. On the territory of the large West Siberian basin in the Late Jurassic and Early Neocomian cyclonian currents have existed (Fig. 2).

North Siberian marine biota differed not only from tethyal but also from lower boreal both by the taxonomic composition and the variety. It was the biota of the boreal type with wide representation of local arctic groups and cosmopolitan organisms, and occasionally with rare lower boreal and subtethys forms. Bivalves were the most variable and numerous representatives from macrobenthos, and from microbenthos - foraminiferes. From other groups often found are Gastropoda, Brachiopoda, Arthropoda (Ostracoda and Malacostraca), warms and sometimes Echinodermate (Crinozoa, Ofiura and scarce Echinozoa) and some other. The building of the benthonic blue-green algae are also known. Pelagic zone has been occupied by half-nectonic and nectonic Ammonitida, Belemnioidea, Teuthida, fishes and sea Reptiles. The remains of the planctonic are represented by Radiolaria, Dinoflagellate and Coccolitophorida. The fossils found in Jurassic-Cretaceous strata in the north of Siberia state all links in the food chains, that have existed in the ecosystems of the north Siberian seas. The group of the producents was represented by phytoplankton (Dinoflagellata, Coccolitophorida and others), the primary consuments concluded Radiolaria, larvas of the marine invertebrates, the presence of which in the composition of zooplancton is out of doubt, foraminifera and others, the secondary consuments included fishes_ plankton-eaters, benthonic filtrates. Tertiary consuments included Ammonitida, Teuthida, Belennitida; consuments of fourth order - Ichtiosaurus and Plesiosaurus; the presence of the destructors: carrion-eaters and bacterias - is supposed proceeding from the necessary reserve of the ecosystem.

Well developed food chains provided circulation of the matter in the ecosystem of the north Siberian seas. However, the structure of biota and distribution of the organisms within the region varied with the time. On the whole, ecosystem of the north Siberian seas was not steady. During Jurassic and Neocomian it changed its state more than 20 times, that expressed in the alternation of the periods of crisises, levellings and differentiations of fauna (Zakharov, Saks, 1980). Periods of differentiation more typical for the Late Jurassic and for the beginning of Neocomian, usually exceeded 2-3 times the duration of crises, which have taken place more often in Early and Middle Jurassic. In spite of changeable structure of biota its taxonomic variety gradually grew and reached its maximum to the end of Jurassic or to the beginning of Cretaceous. With that the definite regularity in the placing of fauna within the basins in the north of Asia is clearly traced: in Early and partly in Middle Jurassic the variety was growing eastwardly and in the Late Jurassic and at the beginning of Neocomian - in the western direction. Such situation is satisfactory explained by paleogeographic features: north Siberian seas in the Early and partly in the Middle Jurassic were closely connected with Paleopacific. The exchange of fauna with the seas of Europe was very difficult (for example in Pliensbachian) and at times (for example in Aalenian and in Bajocian, Fig. 1) it ended.

The territorial neighbourhood of the Early and the Middle Jurassic arctic and tethyc assemblages in western sector of Arctic and North Atlantic is indicative of this. During Late Jurassic and Neocomian the wide connections with the seas in the north of Europe opened, that promoted immigration of lower boreal and subtethycal groups into the north Siberian seas and emigration of the boreal forms to the south. Uneven distribution of the organisms in the north Siberian seas has been determined by their on the bionomical zoning, Bionomical zones were arranged subparallel to the paleoshore, as a rule. Their traces are clearly recorded in the sections of different facies. In the basins with the well worked profile of bottom equilibrium 3-4 bionomical zones corresponding to the conditions of the upper, middle, lower sublittoral and pseude-abyssal are confidently distinguished. Bionomical zone inhabitants were largely unique organisms. Bionomical zones and their parts are particularly reliably recognized on the basis of bottom association by reconstructing paleocatena: the successive line of the bottom associations on the slope of sedimentary basin, reconstructed on the base of study of orycto-cenosis in the outcrops. In the north Siberian Jurassic and Early Cretaceous seas, the more varied benthos associations lived in the shallow water and in moderately-deep biotops. In the direction from paleoshore to the central parts of the basins taxonomic variety diminished and became minimal on the bottom of the pseude-abyssal hollows. For example, in the Late Pliensbachian the upper sublittoral associations were twice as various as lower sublittoral; Callovian - three times and Valanginian upper sublittoral 15-20 times as richer as bottom pseude-abyssal. That regularity was disturbed only in the ecotone zones, where with slight inclinations the variety grows owing to the associations telescoping.

Differenciation of the associations in bionomical zones and their regular placing in the catens were controlled by the same factors, the summary influence of which should rightly be considered as the "effect of shore". Really, away from the paleoshore to the centre of basin in the common case all main characteristics of medium are chaged: types of ground, hydrodynamics, hydrochemistry, temperature, depth, composition and structure biota and other (Zakharov, Yudovni, 1974; Sakharov, Shurygin, 1978, 1979; Shurygin, 1979).

Analysis shows, that each basin comprises peculiar catenas with the special collection of associations (links), the number of which, like a character of ecotons, in the condition of normally falling profile of the bottom was controlled by its inclination (Fig. 3). In the north Siberian seas both, multiunit catenas (of 6-7 links) and with few units (of 2-3 links) ones were distinguished (table, Figs. 4, 5). The more complete catenas were typical for the slightly sloping bottom of the seas of Late Pliensbachian, Late Callovian, Late Volgian-Berriasian and Valanginian times. The catenas with few units were recorded in the Late Aalenian, at the beginning of Bathonian, Oxfordian, in the periods of the unstable profile of the bottom, accompanied by disappearance of the links of the catena or by their telescoping.

As a rule, the catenas with few unites characterized biota under crisis condition. Usually in the composition of such catenas pionner associations prevail which are characterized by qualitative monotony, low population densities of the forming species, little differential trophic structure. Biota, being on the step of differentiation other way, is characterized by many segment catenas consisted from mature climax associations, with the stable trophic structure and with high taxonomic variaty. Gradual complication of the structure of association, growing of the differenciation along the bionomical zones and formation of the most segmental catena to the beginning of Cretaceous we consider to be the evidence of the growing ripeness of the ecosystem of the north Siberian seas during all Jurassic.

REFERENCES

Golbert A.V., Markova L.G., Poliakova I.D., Saks V.N., Teslenko Yu.V. Paleolandscapes of Western Siberia in the Jurassic, Cretaceous and Paleogene. M.: Nauka, 1968, 150 p.

Kaplan M.E. Clacitic Pseudomorphs in the Jurassic and Early Cretaceous Deposits of North Eastern Siberia.- Geol. i. geofiz., 1978, N 12, p. 62-70.

Kemper E., Schmitz H.H. Glendonite-Indikatoren des polarmarinen. Ablagerungsmilieus.- Geol. Rundschau, 1981, Bd. 70, N2, s. 759-773.

Palaeogeography of the Northern Part of the USSR in the Jurassic Novosibirsk: Nauka, 1983, p. 192.

Shurygin B.N. Bivalves and Biofacies in Late Pliensbachian Sea of the Anabar region. In: Environmental Conditions of Mesozoic Sea Boreal Faunas. Novosobirsk: Nauka, p. 82-90.

Zakharov V.A., Radostev I.N. Salinity of Early Cretaceous Sea in North Sibera, According to Palaeobiogeochemical Data.- Geol. i eofiz., 1975, N 2, p. 37-43.

Zakharov V.A., Saks V.N. alaeoecology of the Arctic Basin in the Jurassic and Neocomian. In: Paleontology. Stratigraphy. M., Nauka, 1980, p. 126-132.

Zakharov V.A., Shurygin B.N. Biogeography, Facies and Stratigraphy of the Middle Jurassic in the Soviet Arctic. (by bivalves). Novosibirsk: Nauka, 1978, 340 p.

Zakharov V.A., Shurygin B.N. Jurassic Sea in North Siberia (according to the data of the analysis of bivalves). In: Environmental Conditions of Mesozoic Sea Boreal Faunas. Novosibirsk: Nauka, 1979, p. 56-81.
Zakharov V.A., Yudovni E.G. Conditions of Sedimentation and Existence of Fauna in Early Cretaceous Sea of Khatanga Basin. In: Paleobiogeography of North Eurasia in the Mesozoic. Novosibirsk: Nauka, 1974, p. 127-174.

FIGURE 1. Arctic Sasin in the Aalenian. Conventional signs: dots - land, arrows - directions of the supposed currents (long arrows - surface warm ones, short arrows - bottom cold ones), figures - absolute temperature and salinity (in rectangulars); a thick dot line - the boundary of the Boreal fauna areal, double dot line - the boundary of the arctic fauna areal.

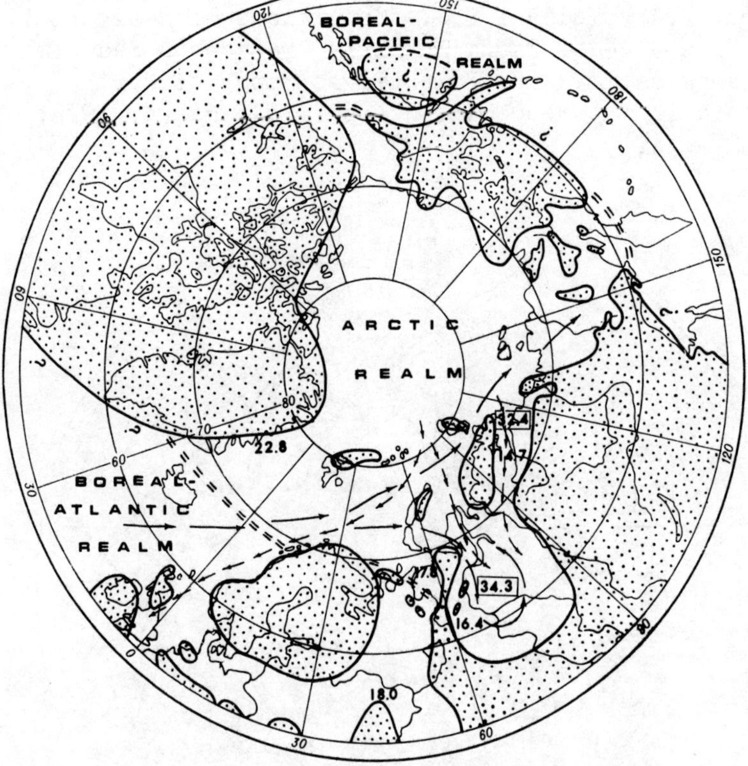

FIGURE 2. Arctic Basin in the Kimeridge.
Conventional signs: see Fig. 1.

(b)

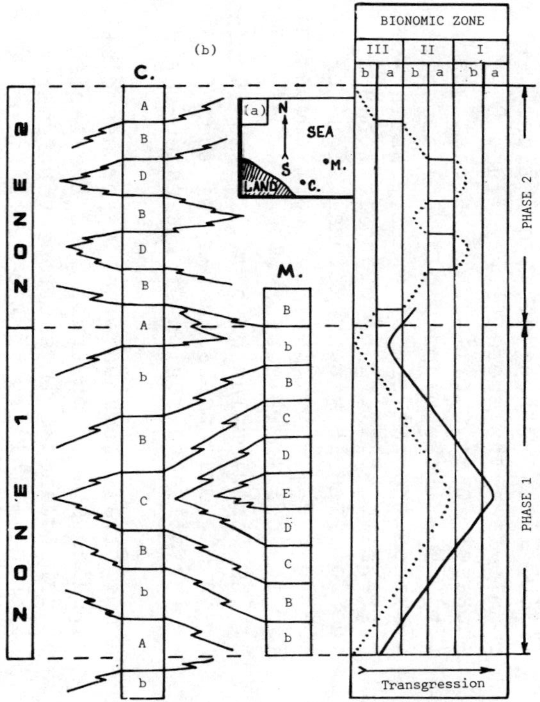

(c)

BIONOMIC ZONE		TIME	PHASE 1	PHASE 2
Upper sublittoral	III	b	assoc. A	assoc. A
		a	assoc. b	
Middle sublittoral	II	b	assoc. B	assoc. B
		a	assoc. C	
Lower sublittoral	I	b	assoc. D	assoc. D
Pseudo-abyssal		a	assoc. E	?

(d)

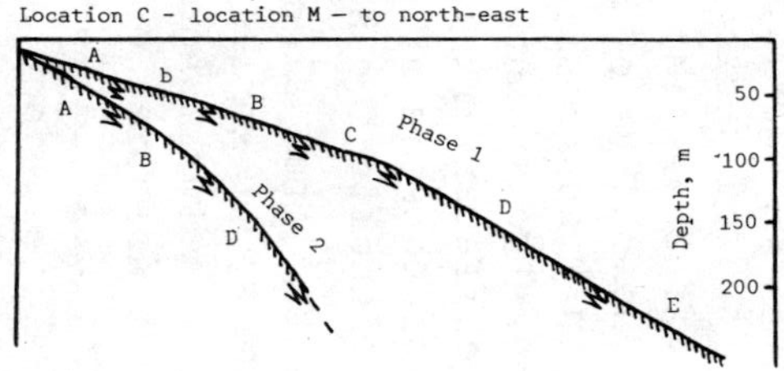

FIGURE 3. Caption overleaf.

FIGURE 3. The scheme of the relief reconstruction of the basin's floor when analysing benthic catens.

a) paleogeographic situation of the studied sections near shore (II) and off shore (M) sections.

b) The change of autochton tanatocoenoses in transgressive-regressive cyclites of two synchronous sections with different facies and the reconstruction of consequent (in time) change of bionomic zones.

Conventional signs: dot lines - isochronoces levels, broken lines - facial transitions; A-E - autochtonous tanatocenoses of shallow water associations (III), those of moderate water associations (II), and those of relatively deep water associations (I); A, B - autochtonoces near shore associations (IIIb, IIb), - those of associations remote from the shore (IIIa, IIa), D - those of lower sublittoral associations (Ib), E - those of pseudoabissal associations (Ia); on the pattern of transgressive - regressive shifting of bionomic zones the dot line is a conventional sign for the near shore section, the continuous one - for the off shore section.

c) The distribution of bivalva associations in bionomic zones (caten) of the investigated region of the paleo-basin.

A-E - bivalva communities, major associations of which are reconstructed on the basis of crictocenoses in sections vertical hatching - falling of the association (bionomic zone) from the caten (on the profile of the bottom).

d) schematic batimetric profile of the investigated part of the paleobasin in various phases.

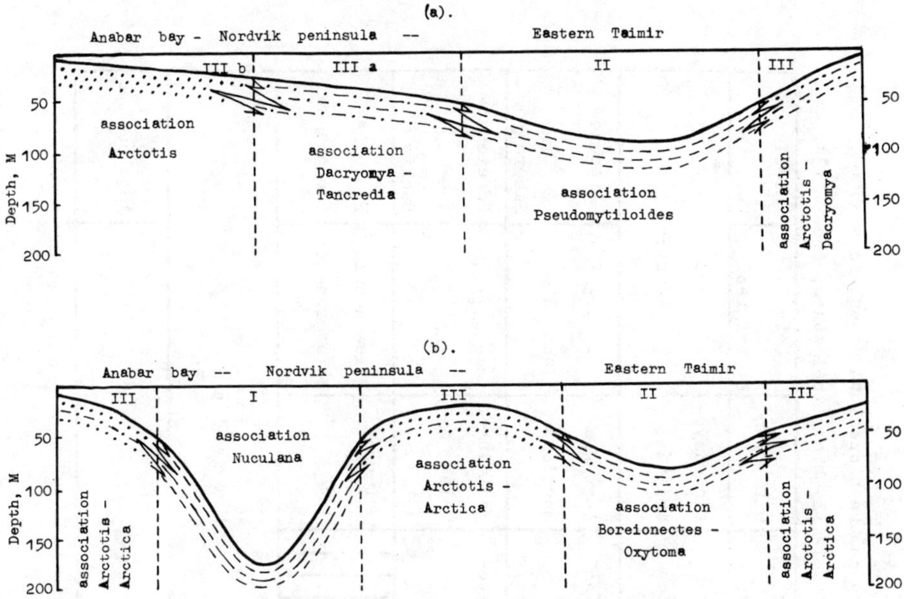

FIGURE 4. Batimetric profile and the caten of benthic Chatanga sea in the Aaleman (a - the early Aaleman, b - the late Aaleman). I-III - bionomic zones. See Fig. 3.

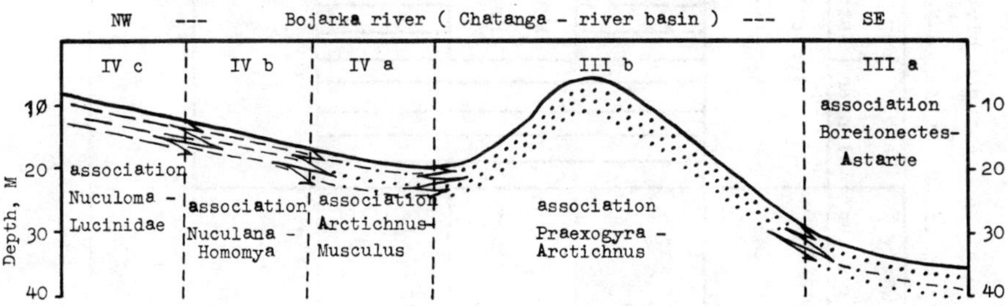

FIGURE 5. Batimetric profile and the caten of benthos of the near shore zone of the south-eastern part (the region of Boiarki) of the Chatanga sea in the early Goterive (Holmolsomites bojarkensis phase), IV - lagoon - marine situations (a - the ones remote from the shore, b, c - near shore ones). Conventional situations, see Fig. 3.

Table. Distribution of the dominant bivalve associations on the bionomical zones in Anabar-Nordvik region of the Jurassic-North-Siberian sea.(I-III- bionomical zones, see fig.3).

	Late Pliensbach.	Early Toarcian	Late Aalenian	Early- Middle Bathonian	Late Callovian	Oxfordian	Late Kimerigian	Middle Volgian
IIIb	Harpax-Myophoria	Liostrea-Meleagri-nella	Arctotis-Arctica		Liostrea-Arctica	Pleuromya-Arctica	Tancredia-Arctica	
IIIa	Meleagrinella-Tancredia				Meleagri-nella-Pleuromya	Astarte-Nuculoma	Buchia-Tancredia	Praeexo-gyra-Pleuromya
IIb	Dacryomya-Radulonec-tites	Dacryomya-Tancredia		Modiolus-Homomya	Grammato-don-Thracia	Modiolus-Meleagri-nella	Entolium Astarte	Boreionec-tes-Buchia
IIa	Modiolus-Homomya					Taimyro-don-Buchia	Palaeonucula-Meleagrinella	Prorokia-Inoceramus
Ib	Malletia-Taimyrodon	?	Malletia - Nuculana					
Ia								Aequipecten